D0164438

Preliminary Version

Earth Algebra

College Algebra with Applications to Environmental Issues

Christopher Schaufele
Kennesaw State College

Nancy Zumoff
Kennesaw State College

HarperCollins*CollegePublishers*

Sponsoring Editor: Anne Kelly
Developmental Editor: Kathy Richmond
Project Editor: Cathy Wacaser
Art Director: Julie Anderson
Text Design: Monotype Composition Company, Inc.
Cover Design: Jess Schaal
Cover Photo: © 1975 Scott Mutter, "Parquet Forest"
Manager, Picture Research and Photography: Nina Page
Production Administrator: Brian Branstetter
Compositor: Monotype Composition Company, Inc.
Printer and Binder: R.R. Donnelly & Sons Company
Cover Printer: The Lehigh Press, Inc.

Earth Algebra: College Algebra with Applications to Environmental Issues,
Preliminary Version

Library of Congress Cataloging-in-Publication Data

Schaufele, Christopher.
Earth Algebra: college algebra with applications to environmental issues/Christopher Schaufele, Nancy Zumoff.—Prelim. version.
p. cm.
Includes index.
ISBN 0–06–500887–1
1. Carbon dioxide—Environmental aspects—Mathematical models.
2. Global warming—Mathematical models. 3. Algebra—Study and teaching. I. Zumoff, Nancy. II. Title.
TD885.C3S32 1992
363.73'87—dc20

92–35839
CIP

93 94 95 9 8 7 6 5 4 3 2

CONTENTS

---

*This section will be provided free to adopters who request it. Contact your local HarperCollins representative for more information.

PREFACE

Approach

This book is a college algebra textbook. The text focuses on modeling real data concerning environmental issues, decision making, reading, writing, and oral reporting. Students work in small groups, and each student is expected to have a graphing calculator. After review of mathematics concepts and skills, much of the work is done in the student groups and the instructor becomes more of a guide than a lecturer. Students write and present oral reports that summarize the work completed for each text section. These components—group work, student reports, and use of the graphing calculator—create a course in which students actively participate.

The environmental focus provides a mass of quantifiable data that is readily available in a context that interests students and faculty. The mathematics used grows out of the need to answer particular questions, some of which are open ended and have many possible answers. Students derive a best model for given data, use it to predict pertinent events, then use it to propose sound limitations for improving environmental conditions.

What excites us about this approach is that students actually use the equations to make decisions about real situations. After we taught our first course at Kennesaw using preliminary notes to this text, we realized that although we have taught a freshman "Decision Math" course for some 15 years, with *Earth Algebra* students were using mathematics to make decisions for the very first time.

The Developmental Story of the Text

In the process of reviewing our departmental course offerings at Kennesaw State College, we were appointed co-chairs of an ad hoc committee and given the charge to do *something* about college algebra. We pondered the question of what makes college algebra boring to so many students. One of our colleagues suggested, "If you want to make a course interesting, then you should study something of interest." To do this we chose to couch most of the standard topics of college algebra in the vitally important issues of the environment. In the fall of 1990 we wrote preliminary notes that were used for pilot units ranging from two to five weeks of a typical quarter. The course has been part of the Kennesaw State College curriculum since fall 1991 and student reaction has been quite favorable.

With the help of extensive reviews by 35 mathematics professors, we revised the project through four drafts of manuscript. The end result is a text that is much improved from our first notes. Class testing at three other schools besides Kennesaw helped us to free the project from our personal teaching styles. We thank Martha Ann Larkin, Southern Utah University; Arthur Sparks, Georgia Southern University; and Lea Campbell, Lamar University-Port Arthur, as well as the Kennesaw mathematics department for taking on the effort and the adventure of using early drafts in their classrooms. Your comments have been very useful to us.

This one-color, "preliminary" version is part of the development of this text. The "first" edition will be available in 1995 and will incorporate additional ideas based on people's experiences using the book—experiences and ideas which we invite you to submit directly to our attention or through HarperCollins.

Distinguishing Features of the Text

This course incorporates the spirit of the *NCTM Evaluation Standards for School Mathematics* in all aspects. This course uses mathematics to study real world problems—in particular, global warming and the greenhouse effect. There is increased awareness of the environment, so most college students enter with at least superficial knowledge of the issues. Through mathematical analysis of real data, students gain a new perspective on mathematics as a tool. Group work, written and oral reports, modeling using mathematics, and the use of graphing calculators makes mathematics a hands-on subject.

Organization of the Text. The text relies on a flexible organization of five parts. Each part opens with an overview of the environmental issues to be examined and the mathematical concepts to be used. Functions traditionally studied in college algebra are used to create models, and students make predictions, decisions, and recommendations about environmental issues based on these models. Each chapter that contains prerequisites ends with "Things to Do" exercises so students can practice before they apply the mathematical skills to the environmental applications. (Students can check some answers in Appendix D.) Each mathematics topic presented is used in an application; there are no extraneous topics. Instructors can choose which parts are to be covered. The Instructor's Guide contains more information about the connections between chapters in *Earth Algebra.*

To complete the environmental applications, the class is divided into groups. Exercises for the groups are indicated in the text by the Group Work icon. Most of these questions are open ended. The students write or present oral reports that summarize the work done for each section of the application.

Part I of *Earth Algebra* discusses the issue of global warming and how it is affected by carbon dioxide build-up. Students are introduced to the ideas of curve fitting and mathematical models. Part II examines three sources of carbon dioxide emissions: automobiles in the United States, U.S. energy consumption, and the destruction of tropical rainforests around the world. For each source, three factors are analyzed and then combined to produce functions describing total carbon emissions from each source. In Part III geometric series are discussed and then used to determine total atmospheric accumulation from each source. Part IV shows how two variables—people and money—affect sources of CO_2 emissions. In Part V alternative energy sources are analyzed through linear programming, then student groups devise plans for decreasing future CO_2 emissions. Each group presents its plan to the class.

Bound-In Supplement. The text contains a Graphing Calculator Manual, which briefly introduces five models—the Casio fx-7700G, the Sharp EL-9200 and EL-9300, and the TI-81 and TI-85. The supplement is especially relevant for students who have had little experience with the graphing calculator.

Besides this bound-in supplement, the text includes appendices that cover the variables used in *Earth Algebra*, a table of conversions for units of measurement, and the derivation of the quadratic formula and a brief discussion of complex numbers. Appendix D allows students to check their answers for the odd-numbered problems in the Things to Do in the mathematics chapters of *Earth*

Algebra. Appendix E contains programs written especially for *Earth Algebra* for use with Casio, Sharp, and Texas Instruments graphing calculators.

Supplement Package to Accompany Earth Algebra

For the Instructor. The *Instructor's Guide* provides background and explanation for the environmental models. It also lists the prerequisite chapters necessary for students to build mathematical models and gives alternative sequencing of topics if an instructor wants a more flexible course outline. Tips for testing or evaluating group work are included. It contains complete worked-out solutions for all models discussed in the text.

The *HarperCollins Test Generator for Mathematics,* available in IBM and Macintosh formats, enables instructors to prepare tests for each of the prerequisite chapters in *Earth Algebra*. Instructors may generate tests in multiple-choice or open-response formats, scramble the order of questions while printing, and produce 25 versions of each test. The system features printed graphics and accurate mathematical symbols. The program also allows instructors to choose problems randomly from a section or problem type or to choose questions manually while viewing them on the screen, with the option to regenerate variables. The editing feature allows instructors to customize the chapter data disks by adding their own problems.

GraphExplorer software, available for IBM and Macintosh hardware, allows students to learn through exploration. With this tool-oriented approach to algebra, students can graph rectangular, conic, polar, and parametric equations, zoom, transform functions, and experiment with families of equations. Students have the option to experiment with different solutions, display multiple representations, and print all work.

Transparencies of selected figures and data from *Earth Algebra* are available for instructors. Transparencies of the keyboards for various graphing calculators (Casio, Sharp, and Texas Instruments) are included for classroom use.

The supplement "Additional Topics to Accompany *Earth Algebra*" might be of interest. *Elementary Algebraic Functions* and *Conic Sections* will be offered to you and your students at no charge should you desire to cover these topics.

For the Student. A *Student's Solution Manual* includes complete, worked-out solutions to selected Things to Do which appear at the end of the prerequisite

A set of *computer-assisted tutorials* offers a self-paced, interactive review of concepts in IBM, Macintosh, and Apple formats. Solutions are given for all examples and exercises as needed. To order, use ISBN 0-06-501286-0 for the IBM package. Order ISBN 0-06-501287-9 for the Macintosh version. For the Apple version order ISBN 0-673-38724-0.

Acknowledgments

Key to the development of this text were the numerous reviewers who took the time and effort to look at various drafts and to give us suggestions and comments. We especially thank the following instructors:

Richard B. Basich, Lakeland Community College
June Bjercke, San Jacinto College
Richelle (Rikki) M. Blair, Lakeland Community College
Bruce Caselman, Westark Community College
Phil DeMarois, William Rainey Harper College
John Dersch, Grand Rapids Community College
Kathleen A. Drude, Millsaps College
Anne Dudley, Glendale Community College
Kay Gura, Ramapo College of New Jersey
William Hemme, St. Petersburg Junior College
Bruce Hoelter, Raritan Valley Community College
Roseanne S. Hofmann, Montgomery Community College
Ingrid Holzner, University of Wisconsin-Milwaukee
Joe Klerlein, Western Carolina University
Raymond Lee, Utica College of Syracuse University
Rowan Lindley, Westchester Community College
LuAnn Malik, Community College of Aurora
Wendy Metzger, Palomar College
Judith Ann Miller, Delta College
Margaret Barnes Morrison, San Jacinto College
Jack R. Porter, University of Kansas
David Price, Tarrant County Junior College
David Reichard, Charles County Community College
Rosalind Reichard, Elon College
A. Allan Riveland, Washburn University of Topeka

A. Allan Riveland, Washburn University of Topeka
David Royster, University of North Carolina-Charlotte
William A. Sanders, University of Wisconsin-Platteville
Don Shriner, Frostburg State University
Loretta J. Robb-Thielman, University of Wisconsin-Stout
Jon D. Weerts, Triton College
Joseph C. Witkowski, Keene State College

We thank all of our colleagues and staff at Kennesaw State College, in particular, we acknowledge and are deeply appreciative to: Pamela Drummond for the project evaluation; Marlene Sims and Stanley Sims for advice in the early development stages; Marian Fox for suggestions on the manuscript; student assistants Pemle Ennis, Sherry Hix, Jill Roberts, and Debbie Beck for manuscript preparation; and Dean Herbert L. Davis for his early and continued support. We are especially grateful to our department chair, Tina H. Straley, without whose support this entire project would have been impossible.

Also we wish to express thanks to: Jack Pritchard and Anne Kelly for being so venturesome as to publish this book; Kathy Richmond for her patience and editorial assistance; John Kenelly, for his curriculum guidance and calculator expertise; and Ben Fusaro, Chair of the MAA Committee on Mathematics and the Environment, for being a true friend to the earth.

Finally, a special thanks to Pauline, to Lanier, and to Nancy Leigh for their understanding, tolerance, and encouragement.

This project is supported by generous grants from the National Science Foundation and the U.S. Department of Education: Fund for the Improvement of Post-Secondary Education. We are proud and grateful.

Christopher Schaufele
Nancy Zumoff

This work was supported by NSF Grant Number USE–9150624 and U.S. Department of Education: FIPSE Grant Number P116B10601.

TO THE STUDENT

> "If you want to make a course interesting, then you should study something of interest."
>
> M. Sims, March 1990

These simple but deceptively wise words were delivered at the first meeting of an ad hoc committee at Kennesaw State College. The charge to this committee was the following: "Do something about college algebra." Although our Department Chair claims that we volunteered, we, in actuality, were appointed co-chairs of this committee. This appointment was an immediate and direct response to this public statement: "We are wasting our time and the students' time teaching our existing college algebra course."

The opening profundity of Ms. Sims could very well have been the impetus for this book. What, indeed, could be done to make college algebra interesting? And, what makes college algebra boring to practically every student on this planet? The answers to the latter question were easily determined by a review of questions perennially posed by its students. Here are a few familiar ones:

1. "What's this stuff good for?" (in response to most anything);
2. "Who cares?" (in response to thought provoking word problems, such as "Train A leaves New York...," or "Sally is twice as old as John...");
3. "When will I ever have to do this again in my life?" (in response to simplification of a complex fraction that only Rube Goldberg could have designed);

and lastly, our favorite:

4. "Is x always equal to 2?" (in response to having solved a hideous equation involving roots of rational expressions).

There are answers, of course, to all these. In reverse order:

4. "Yes."

3. "I'm not sure."

2. "*You* should, if you want to pass this course."

1. "Designing electrical circuits," "constructing bridges," "putting a woman on the moon," etc., etc., etc.

The first question probably encompasses all the others (except possibly number 4), so we briefly observe that all the answers provided to this query are true but rather inadequate in yielding a meaningful link between factoring and space walking to any student at the beginning college level.

One of our principal goals, perhaps we should say dreams, is to forever lay question 1 to rest, at least among students of *Earth Algebra*. In an attempt to provide the interest, we have couched college algebra in the vitally important issues of the environment. Neither train A, nor Sally's age, will be of concern herein. Real data about real things are provided in this text, and models are derived to fit. Models use relatively simple algebraic equations, such as linear, quadratic, and rational; also exponential and logarithmic functions are summoned when appropriate. We define a "best" model for a given set of data, and after its derivation, students use it to predict relevant events, then to impose reasonable societal restrictions for improvement of future environmental conditions. The equations are actually used to make decisions about real-world situations. After completion of the first pilot segment of this course, both of us commented that we have been teaching freshman mathematics courses entitled "Decision Math" for some fifteen years now, and this was the first time our students had ever used mathematics to make a decision.

It is our intention that the majority of the material in this text be studied in small groups, ranging from three or four students each. We have found that this stimulates responsibility and confidence. Students' written reports and oral presentations incorporated into the studies place real meaning to "$x = 2$." Two what? Two decades? Two tons of carbon dioxide? And whatever "two" is, how is it relevant?

The text is designed for study with the aid of graphing calculators. These mercifully remove the deep, dark tedium and drudgery involved in manipulation of demonic algebraic expressions.

Earth Algebra is intended for study in a beginning level college algebra course. Most of the standard topics usually covered in a traditional college algebra course are included, although most manipulation of algebraic expressions and graphing of more sophisticated equations are handled with the calculator. However, with the aid of the graphing calculator, we are able to include other topics which may not normally have been covered in the traditional course. A warning: students wishing to enter a standard calculus course may need additional preparation.

Overall, it is our sincere wish that both the student and the instructor find the reading of this book enjoyable and educational. And to the student: may you learn a little mathematics—and what it's good for—along the way.

Finally, this project is supported by generous grants from the National Science Foundation and the U.S. Department of Education: Fund for the Improvement of Post Secondary Education. The authors are proud and grateful.

Christopher Schaufele,
Professor of Mathematics
Kennesaw State College

Nancy Zumoff
Professor of Mathematics
Kennesaw State College

PART I

Carbon Dioxide Concentration and Global Warming

INTRODUCTION

It was a wintry Thursday afternoon, sunny, but 15° with 30 mph winds, and the furnace was not working in the mathematics building. The old professor had just finished teaching her morning calculus class. She gathered up her notes, and instead of returning to her cold office, walked straight through the front doors, and across the campus quadrangle to the parking lot where her car was waiting in the sunlight. She opened the door on the passenger side, slowly sat down in the seat, thumbed through her notes a minute, then comfortably remained there for the next hour preparing for her upcoming analysis class.

What the old professor was doing was enjoying the greenhouse effect. Even though it was below freezing, the sun had warmed her closed automobile to a comfortable temperature. And that's what the greenhouse effect is: light passes through some enclosing barrier, heat gets trapped within, and it warms the enclosure.

The earth itself is surrounded by a sort of blanket. This blanket is composed of natural gases and traps the heat sent down from the sun, holds it around the surface of our planet, and keeps the air warm. These gases are known as "the greenhouse gases" because the blanket they form works like a greenhouse: it allows sunlight to pass through, but holds in a certain amount of heat.

The proportion of these gases has remained about the same from the end of the last ice age until the beginning of industrial times. As a result, our climate has remained relatively stable. The gases which are the major contributors to the greenhouse effect are, in order of quantity: carbon dioxide (50%), chlorofluorocarbon and other halocarbons (20%), methane (16%), ozone (8%), and nitrous oxide (6%).

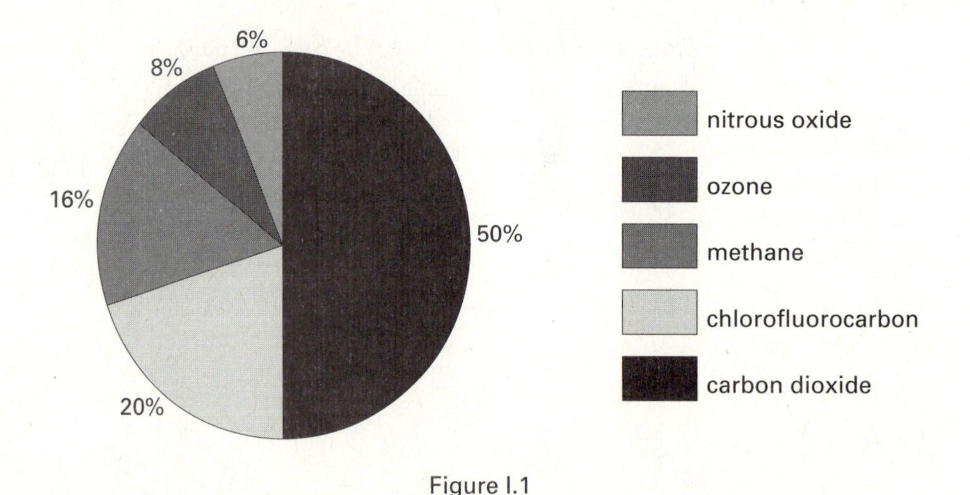

Figure I.1

Carbon dioxide, CO_2, results from the burning of fossil fuels and deforestation; chlorofluorocarbon, or CFCs, come from air conditioners, refrigerants, foams, and aerosol propellants; methane is produced by swamps, landfills, cattle and other livestock; ozone is a result of fossil fuel burning; and finally, nitrous oxide comes from agricultural chemicals fertilizer, and also from the burning of fossil fuels.

The thing to note here is that the climate on our planet is O.K.—as long as these gases maintain their present proportions. But the world population is increasing rapidly. And all these new people need shelter, need to keep warm or cool, cook food and then eat it, and travel from here to there. Many of these new people want Styrofoam coolers, aerosol spray cans of this and that, and the convenience of the myriad of disposable items available at the stores. Then, of course, there need to be lots of new stores, or malls, to sell all these things to all these people. And many of these people use plastic or paper bags to carry their packaged purchases out to their cars. Then there's all the left-over materials, and the no-longer-of-use, or -interest, products which wind up in a landfill.

All these things produce greenhouse gases. So with this increase in production of these gases, it might be suspected that the "normal" proportions present in the atmosphere will be affected. More and more greenhouse gases could make the blanket thicker and thicker, thereby trapping more and more heat at the surface of the earth. And what we get is an increase in average world temperature, which is known as global warming.

In the past one hundred years, the average global temperature has increased 1° Fahrenheit. One degree increase over a century may, at first glance, seem insignificant, but note this fact: from the last ice age 18,000 years ago, the earth has warmed by 9° F, or .05° F per century. The present rate of increase is twenty times that of the average rate over the past 180 centuries. That's significant! And, even scarier, the six warmest years in the history of recorded temperature are, in decreasing order of degrees, 1990, 1988, 1983, 1987, 1984, and 1989.

If this trend continues, the eventual effect would be the end of life as we know it on planet earth. One obvious consequence is a change in climates, which would seriously affect farming lands and water supplies. Also, water would expand from the heat and polar ice caps would start melting, causing oceans to rise and cover beaches and other low country. If the global temperature continues to rise at its present rate there will not be time for evolutionary adaptation, thus causing mass extinction of plant and animal species. All these things pose a serious threat to the existence of us humans as well.

Hopefully, most of you are already well aware of the information you have just read, and are doing your environmental duty to save the planet. This book is not intended to provide you with much more information about the causes and effects of global warming. What this book is intended to do is to use concepts from college algebra to study greenhouse gas emission and use these concepts to predict the future. Now this may seem weird to you at this point, i.e., how to use mathematics to study about environmental problems. Well, we're going to tell you briefly how it's done—but only briefly—and then we're going to show you. Everybody knows that the only way to learn something is to do it, and that brings to mind the old adage: "Math is not a spectator sport."

In order to study physical, or social, or whatever, situations, mathematicians and statisticians build a "model" which looks as much like the situation as possible. The kind of model we use in these notes consists of equations and formulae and things of this nature. Here's how we will get these equations and formulae: we have access to historical data which can be plotted on a coordinate system; we find an equation whose graph comes close to all the points determined by the given information. This equation should provide a fairly good approximation to what actually occurred in the past, and can then be used to predict what could be expected to happen in the future, should things continue as they have. Of course, some curves will come closer to the plotted points than others, and we will need to decide which curve is best to use, i.e., which equation best approximates the real data. The types of equations we have at our beck and call are linear, quadratic, logarithmic, and exponential; so, given our real data, we first need to decide which type of equation has its graph most shaped like the plotted points. Then we need to determine how to find the best one of these to use i.e., the one which approximates the actual figures the closest.

The procedure described above is called *curve fitting*, and the collection of all of the equations needed to describe a situation is called the *model*.

To the student: the models you will use and build in this course are quite simplistic. This is necessary because of the scope of the course. You should be aware that in reality, many, many other variables, social and political as well as physical, become involved. We have, however, based all models on real data, and most of them prove to be pretty consistent with published predictions.

The mathematical models in *Earth Algebra* are based on current and past trends. Prediction of future events using these models is based on the assumption that these trends will continue. It may not be the case that these trends actually continue; therefore you must understand that your predictions may not come true. Many things can happen in the future which affect these trends, thus changing the outcome.

Also, a model may be accurate for a short period past the present date, but certainly over a long period of time, many other unforeseen events will affect the accuracy of the predictions.

In Part I of this book, you will use the mathematical concepts of functions, linear functions, and composite functions to study atmospheric carbon dioxide concentration and its effect on average global temperature and ocean level.

No more discussion—on to the real thing!

CHAPTER 1

Functions

1.1 NOTATION AND DEFINITIONS

We'll try to keep this section simple. For the most part, you'll need to deal with functional notation when you do *Earth Algebra*, so we'll start with that. Almost all of the *Earth Algebra* functions are defined with algebraic expressions, or logarithmic or exponential expressions, or combinations of these. For example,

$$2x + 7$$

defines a function of the *variable* x. If you substitute your favorite number for x and do the arithmetic, you'll get a number for an answer. If your favorite number is $x = 5$, then substitute 5 for x:

$$2(5) + 7 = 17$$

is the answer.

A really "loose" way to define function is this: a function assigns to each number x a unique number y.

In the above example, the function defined by $2x + 7$ assigns to the number $x = 5$ the number $y = 17$. It assigns to the number $x = 0$ the number $y = 7$, etc., etc.

Note: Chapter 1 is a prerequisite for Chapters 2 and 3.

Just like people, functions need names. Name the one defined by $2x + 7$ by the letter f. Now f is a function of the variable x, so functional notation is

$$f(x) = 2x + 7.$$

Read the thing on the left of the equal marks: "f of x." It works like this: if you substitute 5 for x, then you write

$$f(5) = 2(5) + 7 = 17,$$

or if $x = 0$,

$$f(0) = 2(0) + 7 = 7.$$

You say for these operations: "$f(x)$ is evaluated at 5", or "at 0," or, abbreviated, "f of 5" or "f of 0."

Whatever you substitute for x goes in the parentheses in the functional notation: $f(-1) = 2(-1) + 7 = 5$. If you substitute a cat for x, then

$$f(\quad) = 2(\quad) + 7.$$

Got it? Some people think that $f(x)$ means multiply f times x, but it doesn't. "f" is the name of the function and x is its variable.

A couple more quick examples of functions are:

1. $f(x) = x^2 - 3x + 1$
2. $g(x) = \dfrac{1}{x-3}$

Evaluate $f(x)$ at 2: $f(2) = (2)^2 - 3(2) + 1 = -1$; evaluate $g(x)$ at 7: $g(7) = \dfrac{1}{7-3} = \dfrac{1}{4}$.

It isn't hard to evaluate functions on your calculator, but each kind does it a little differently. First you have to tell the calculator what your function is, that is you must enter the expression for your function (which might be called $y_1, y_2, \ldots$, or $f_1, f_2, \ldots$). Then assign the value to x, press the key for your function and enter. For example, to evaluate $2x + 3$ at -5, you would enter the expression

$$2x + 3$$

for the function y_1 (or f_1), assign -5 to x, press y_1 (or f_1), and enter. Your screen will show -7. To find out more, read your manual, or ask your instructor or a knowledgeable friend.

1.2 DOMAINS AND RANGES

Try this: evaluate the above $g(x)$ at 3. Big problem. Division by zero is illegal. This means that you can't just substitute any number into a function. The function $g(x)$ is not defined when $x = 3$, so you may have to restrict the numbers which can be used to evaluate a particular function. The only real problem with evaluating $g(x)$ occurs when the denominator equals 0, i.e. when $x = 3$, so $g(x)$ is defined for all numbers except 3. In general, the domain of a function consists of all numbers at which the function is defined. Thus the domain of $g(x)$ consists of all numbers except $x = 3$.

What about the domain of $f(x) = x^2 - 3x + 1$? No problem here with substituting any x, so its domain consists of all numbers.

A kind of different situation is presented by the function

$$h(x) = \sqrt{x}.$$

No negative numbers can be substituted for x because you can't take the square root of a negative number and get a real number. In *Earth Algebra*, imaginary numbers will not be considered. (See Appendix C for imaginary numbers, or complex numbers.) But, anything else is O.K. to substitute in for x, so the domain of $h(x)$ is the set of all real numbers $x \geq 0$. See what happens if you try to evaluate this function at $x = -3$ using your calculator.

There are also practical considerations to take into account when worrying about the domain of a function. All of the functions of *Earth Algebra* have practical purpose in their lives. For example, suppose that $F(w)$ defines the number F of catfish which live in a certain North Georgia lake as a function of the number w of gallons of toxic waste dumped into a creek upstream of the lake by a certain chemical factory. ($F(w)$ could be defined by some sort of mathematical expression, but that doesn't matter now. But whatever defines it, you can bet that it is decreasing; i.e., the larger w is, the smaller F gets!) The variable w counts gallons of toxic waste so it would make no sense to say $w = -10$, right? "Negative ten gallons of waste were dumped today." Your friends would think you are crazy. So the domain of $F(w)$ would be all: $w \geq 0$.

There are two things to consider when determining the domain of a function:

1. Where is the mathematical expression defined?

2. What are the practical limitations?

EXAMPLE 1.1

Determine the domain of $f(x) = 2x + 1$.

Anything can be substituted for x. Domain = all real numbers.

EXAMPLE 1.2

Determine the domain of $f(x) = \sqrt{x-1}$.

Whatever's under the radical must be non-negative; i.e., $x - 1 \geq 0$; thus $x \geq 1$. Domain = all real numbers $x \geq 1$.

EXAMPLE 1.3

$J(p)$ defines the unemployment rate J (for jobs) as a function of gross national product p. The gross national product is another counting variable, it is the monetary value of all goods produced in the country. Hence $p \geq 0$, and the domain of $J(p)$ would be all real numbers $p \geq 0$.

Up until this point, we've only been talking about the numbers which can be substituted into the function (domain), but what about the answers you get after you substitute? All the answers that you get when you substitute all the domain numbers into the function comprise what is known as the range of the function.

If $f(x) = x^2$, then $f(2) = 4$, $f(-1) = 1$, $f(7) = 49$, etc., etc. So the numbers 4, 1, 49 are all in the range of $f(x)$; of course these are not all the range numbers; you could plug in numbers all day and never get all the range of $f(x)$. There's a nicer way to see what the range is graphically; we'll talk about that later. But do note this now: -4 is not in the range because $x^2 \geq 0$ always. As a matter of fact, no negative number is in the range.

This is enough about range for now—more later.

1.3 GRAPHS OF FUNCTIONS

We've heard lots of students say that they hate graphs, but they shouldn't. Graphs are useful; they are pictures of functions. Instead of dealing with some complicated mathematical expression which defines a function, you can graph it and see more about what's going on.

If $f(x)$ is your function, then first set $y = f(x)$. If $f(x)$ is defined by some expression, then what you now have is an equation in two variables x and y. x is called the *independent variable* (because that's the one you substitute the numbers for), and y is called the *dependent variable* (because y is the answer and it depends on what x is). The graph of the function $f(x)$ is sketched in the plane formed by an (x, y) coordinate system. This consists of a horizontal axis for x, or whatever the independent variable is, and a vertical axis for y, or whatever the dependent variable is. Each of these axes is a copy of the real line; they intersect at zero on each, positive numbers are to the right and up; negative numbers are to the left and down. Points in this plane are located by a pair of numbers (a, b); the first one, a, comes from the horizontal axis, and the second, b, comes from the vertical axis. See Figure 1.1.

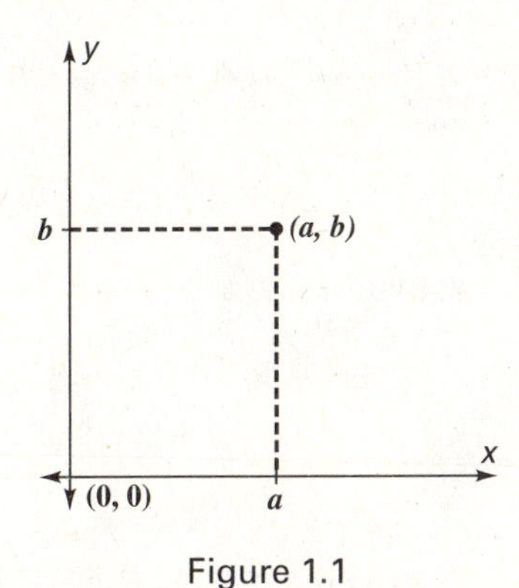

Figure 1.1

If $x = a$, and $b = f(a)$, then (a, b) is a point on the graph of $f(x)$. The *graph* of $f(x)$ consists of all points (a, b) such that $b = f(a)$, and a varies over the entire domain.

Remember when you first learned how to graph a function, or equation, you'd substitute three or four numbers in for x, find the corresponding y value, plot your points and connect the dots? This may not give a very good picture of your function because you left out so many domain numbers. You ask how could anyone possibly substitute every one of that infinity of numbers in the domain. Of course no one can. But there are two ways to get the graph without doing that: one way is to learn calculus; the other is to use your calculator. As you study each type of function in *Earth Algebra*, we'll tell you about its particular intricacies—what's important about each—then discuss how to see these important things on your calculator screen. Graphing with the calculator can be fun. And there'll be plenty for you to practice with when the time comes.

One final note on functional notation: none of the functions of *Earth Algebra* use the variables x and y. They are named according to the purpose they serve. For example, suppose you wanted to write a function which could predict the number of acres of rain forest which would be remaining on this planet in any given year. What would you name your function? How about RF? And what would RF be a function of? How about t (for time)? So your function would be $RF(t)$ where RF = number of acres of rain forest in year t. And then your graph would be on a (t, RF) coordinate system. See how that works?

Finally, here is some practice for you.

1.4 THINGS TO DO (ALIAS EXERCISES)

For Exercises 1–10, evaluate the given function at the indicated values, and find the largest possible domain for each. Try evaluating the first three functions "by hand," and for the others, let your calculator do the work.

1. $f(x) = 5 - 2x$; evaluate $f(x)$ at 2, –3, 1.01, and $-\frac{1}{2}$.

2. $g(x) = 2x^2 - 5x + 17$; evaluate $g(x)$ at 2, –3, 1.01, and $-\frac{1}{2}$.

3. $h(x) = \dfrac{3x^2 - 5}{4x - 7}$; evaluate $h(x)$ at 2, –3, 1.01, and $-\frac{1}{2}$.

4. $F(t) = 517 - 2.14t$; evaluate $F(t)$ at 2, –3.15, 0.000051, $\frac{7}{5}$, and $-\frac{1}{2}$.

5. $T(x) = .005x - 120$; evaluate at $x = 2.4$, 24, 240, 2400, and 24000.

6. $G(x) = .002x^2 - 5.13x + 729$; evaluate $G(x)$ at 2, –31, 121.201, $\frac{2}{3}$, and $-\frac{1}{2}$.

7. $H(x) = \dfrac{3x^2 - 5}{2x - 11}$; evaluate $H(x)$ at 2, –3.3, 11.001, and 5.5.

8. $S(u) = \dfrac{(.005u - 20.2u^2 + 9.1)(10.9 + 2.3u)}{12u - 2.1}$; evaluate $S(u)$ at 0, 50, 100, and 5000.

If you want to do something different, try this on your calculator: enter each of the parenthetical expressions as a y_1, y_2, and y_3, then let $y_4 = S(u)$, the appropriate combination of the three pieces.

9. $R(s) = \sqrt{63 - 2.1s}$; evaluate at $s = 0, -2, 30, -1.52$, and 50.

10. $P(x) = 150{,}000x(12 - .0005x)$; evaluate at $x = 10, 100, .001$, and 108.

CHAPTER 2

Linear Functions

2.1 SLOPE-INTERCEPT EQUATION

The graph of a linear function is a straight line. A linear function looks like $f(x) = mx + b$, where m and b are constants. These constants are very significant: the number m is the *slope* of the line, and the number b is its *y-intercept.* (The y-intercept of any graph is the place where it crosses the y-axis, and the *x-intercept* is the place where it crosses the x-axis.)

The y-intercept is found by setting $x = 0$ (any point on the y-axis has $x = 0$) and then determining what y is.

If $x = 0$, then $f(0) = m \cdot 0 + b = b$, so $y = b$ is the y-intercept, just as we said!

Now, to see the significance of the slope, set $x = 1$ and substitute:

$$f(1) = m \cdot 1 + b = m + b.$$

You have this information:

$$\text{when } x = 0,\ y = b;$$

$$\text{when } x = 1,\ y = m + b.$$

See Figure 2.1.

Note: Chapter 2 is a prerequisite for Chapter 3.

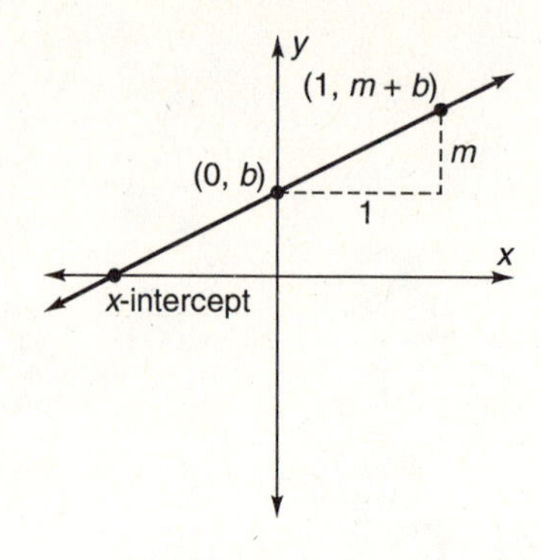

Figure 2.1

This says that a one-unit change in x (from 0 to 1) results in an m-unit change in y (from b to $m + b$). The horizontal axis is the x-axis, so any change in x represents a horizontal change; the vertical axis is for y, so changes in y are vertical changes. A one-unit horizontal change in the graph of $y = mx + b$ results in an m unit vertical change. This is why m is called the *slope* of the line; it tells how steep it is. If you look at any two points on your line, and measure the horizontal change and the vertical change from one point to the other, then this slope is

$$m = (\text{vertical change}) \div (\text{horizontal change}).$$

Look back at the original function

$$f(x) = mx + b,$$

or the equation

$$y = mx + b.$$

This is called the *slope-intercept equation* for a line (here's a question: why?). This is a nice equation because it's simple and gives lots of information about the line.

Some particular examples are below.

EXAMPLE 2.1

Determine the slope and y-intercept of $y = 2x + 1$.

Slope = 2, y-intercept = (0, 1).

See Figure 2.2.

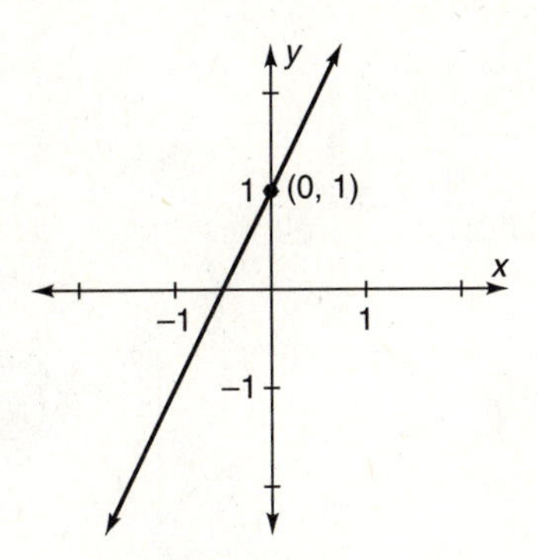

Figure 2.2

EXAMPLE 2.2

Determine the slope and y-intercept of $y - 1 = -2(x - 3) + 7$.

First write in slope-intercept form.

$$y = -2x + 14.$$

Slope $= -2$, y-intercept $= (0, 14)$. See Figure 2.3.

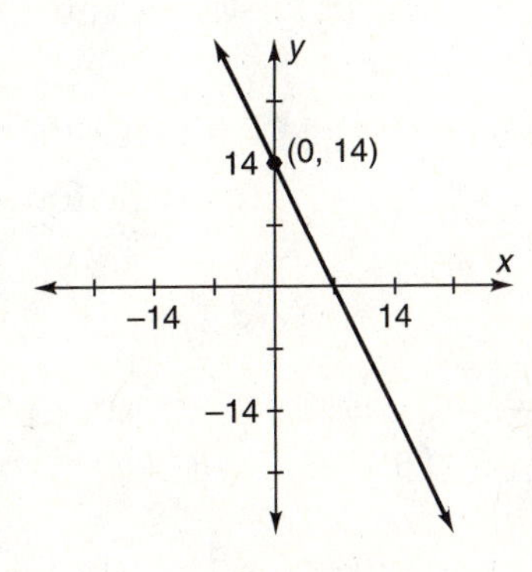

Figure 2.3

EXAMPLE 2.3

Determine the slope and y-intercept of $2y = 3x + 1$.

Careful, the slope is not 3 nor is the y-intercept (0, 1). Write in slope-intercept form; i.e., solve for y:

$$y = \frac{3}{2}x + \frac{1}{2}.$$

Now the slope is the coefficient of x, and the y-intercept is the constant at the end.

$$\text{Slope} = \frac{3}{2},\ y\text{-intercept} = (0, \frac{1}{2}).$$ See Figure 2.4.

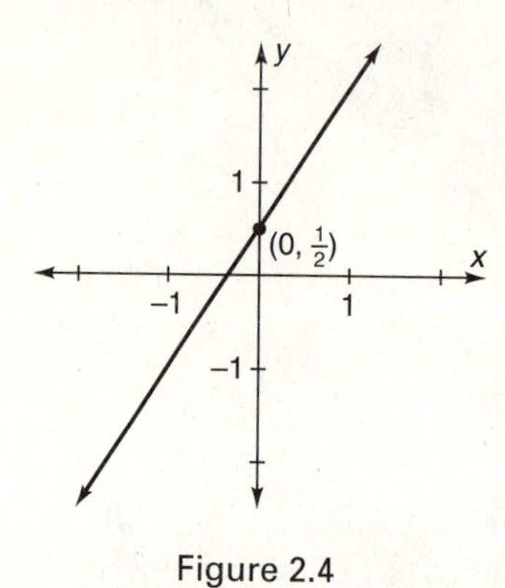

Figure 2.4

2.2 SLOPES

Let's get numerical! First, remember that the slope is the ratio of the vertical change to the horizontal change between two points on the graph of a line. If the two points are (x_1, y_1) and (x_2, y_2), the horizontal change is $x_2 - x_1$ and the vertical change is $y_2 - y_1$. Hence the formula

$$m = \frac{y_2 - y_1}{x_2 - x_1}$$

gives you a way to determine the slope of a line if you know two points on that line. See Figure 2.5.

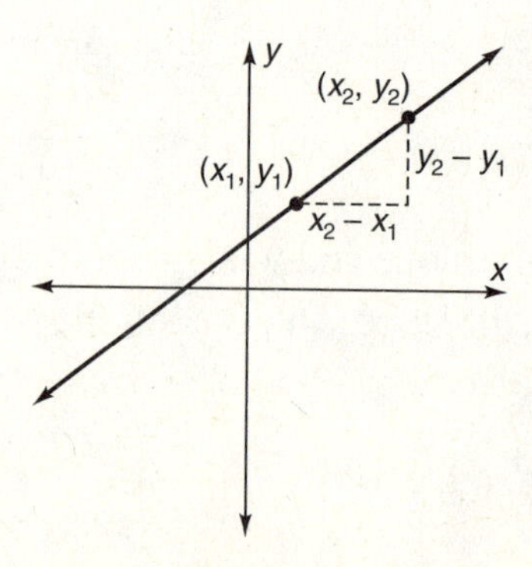

Figure 2.5

What if you needed to know the slope of the line which goes through the points (1, 2) and (3, 4)? Having mastered college algebra, you could very quickly answer, "It's quite simple. The slope is 1." But then suppose "minuses" got involved and the points changed to (–1, 2) and (3, –4)? But negatives wouldn't bother you in the least. You reply, "That slope is

$$m = \frac{(-4)-2}{3-(-1)},$$

which is $\frac{-6}{4}$, and that's $-\frac{3}{2}$."

There's an easy way to get the graph of a linear function using its slope. It's what we call "Doing the three-second graph." Suppose you know a point on a line and the slope of that line. Then think of slope in terms of ratio of vertical change to horizontal change to draw this graph. An example or two illustrate this concept best.

EXAMPLE 2.4

Graph the line which passes through (2, 5) and has slope 4.

Write $m = 4$ as a fraction,

$$m = 4 = \frac{4}{1};$$

then the vertical change is 4 and the horizontal change is 1. Your starting point is (2, 5). Move vertically 4 units up, then horizontally 1 unit to the right. This determines another point on this line; that point is (3, 9). You can count points on the coordinate system to get this point or you can add to get $(2+1,\ 5+4)$. Now, simply draw the line through these points. See Figure 2.6.

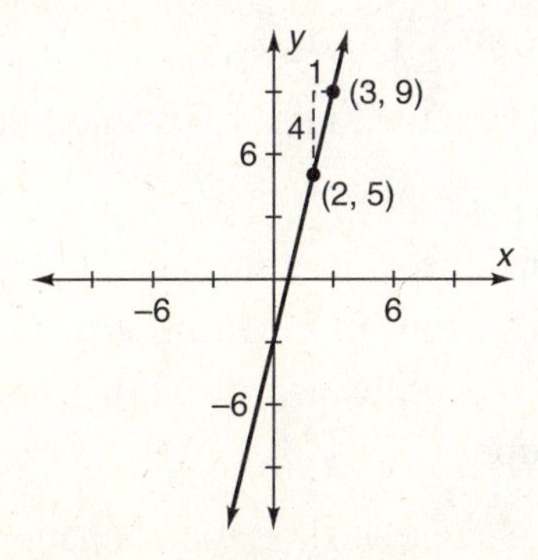

Figure 2.6

EXAMPLE 2.5

Graph the line which passes through (1, 2) and has slope $\frac{-3}{5}$.

Think of the slope as being –3 divided by 5; write the slope as

$$m = \frac{-3}{5},$$

so the horizontal change is 5 and the vertical change is –3. The negative in front of the 3 means that your vertical change is downward. Start at (1, 2), go down 3 units, then right 5 units. The new point is $(1 + 5,\ 2 - 3) = (6, -1)$. Draw the line through (1, 2) and (6, –1) and you've got it! See Figure 2.7.

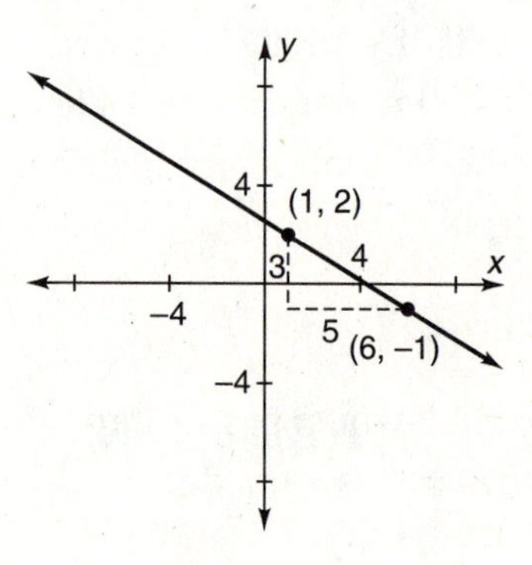

Figure 2.7

Before we move on, here are some useful pieces of information about slopes.

- If $m > 0$ the linear function is increasing, i.e., the graph is rising from left to right.
- If $m < 0$, then the linear function is decreasing, or its graph is falling as you go from left to right.
- If $m = 0$, then the line is horizontal; the linear function is constant. See Figure 2.8.

There's this one other thing about slopes and lines which you won't be concerned much about in *Earth Algebra*, but we'll tell you anyway—just so your education is complete: vertical lines have no slope—their slopes are undefined. Since slope is

(vertical change) ÷ (horizontal change),

a vertical line has zero horizontal change; so you'd have to divide by zero, which is undefined. Finally, the equation of a vertical line looks like $x = a$, where a is constant. See Figure 2.9.

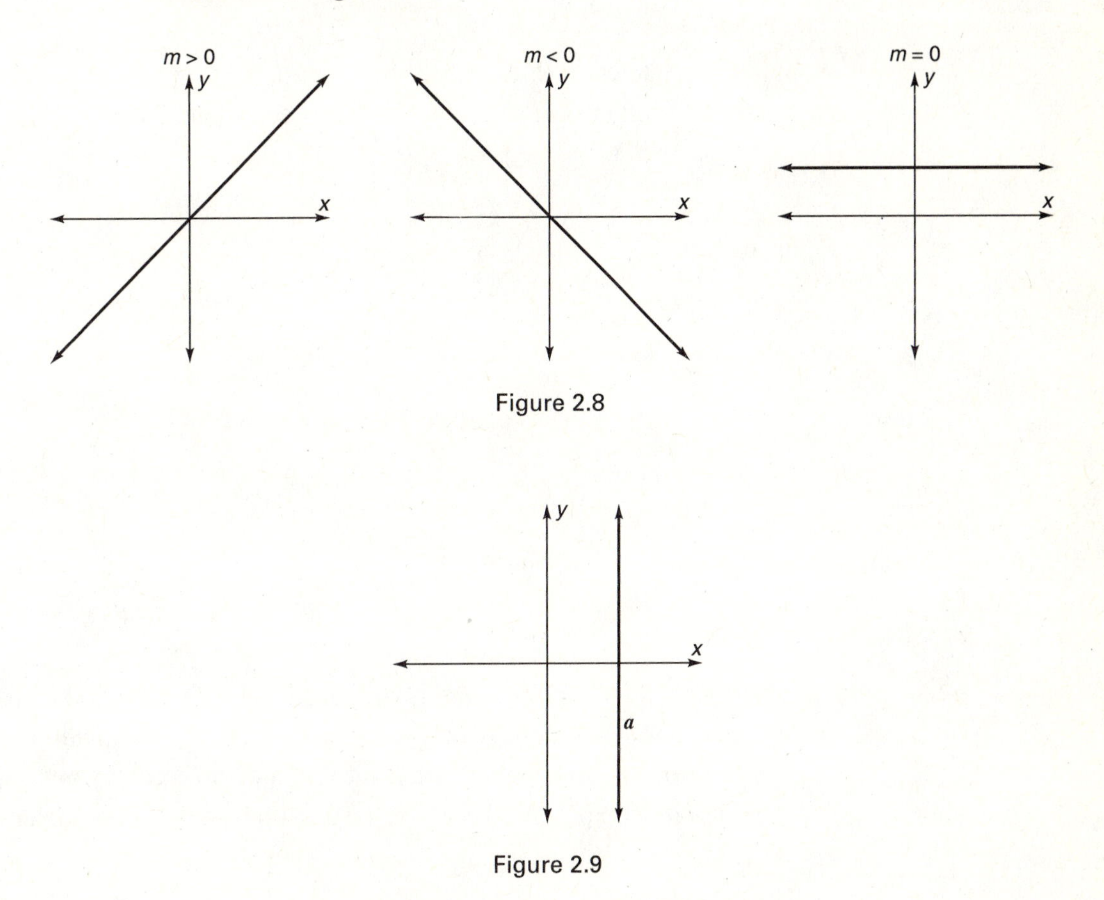

Figure 2.8

Figure 2.9

2.3 GRAPHS

Linear functions have, of course, the easiest kind of graph. All you need are two points, and any two points uniquely determine a line. If you've carefully read the sections which come before this, you already know one way to graph a linear function, the slope-intercept way: first, write your linear equation in the form $y = mx + b$; then start at the y-intercept and think of the slope as

$$(\text{vertical change}) \div (\text{horizontal change}).$$

An example is:

$$y = -2x + 1. \text{ (Note: } m = \frac{-2}{1}.)$$

Start at the y-intercept (0, 1), move right 1 unit, then move down 2 units to get your second point (1, –1) and draw the line. See Figure 2.10.

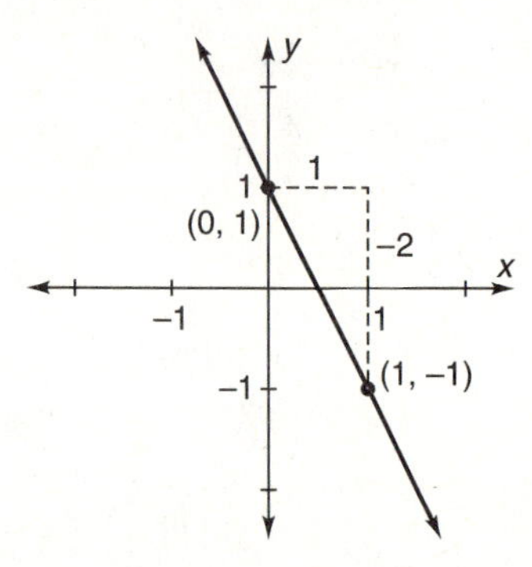

Figure 2.10

Another way to graph a line is to determine both the y-intercept and the x-intercept. The y-intercept is easy to determine when the equation is in slope intercept form $y = mx + b$; it's $(0, b)$. But to get the x-intercept, you have to work a little—not much, though. What you have to do is to solve the linear equation

$$0 = mx + b$$

(this is a result of setting $y = 0$). An example of this follows.

EXAMPLE 2.6

Graph the line $y = 3x + 6$ by finding its x- and y-intercepts.

The y-intercept is (0, 6). For the x-intercept, take $y = 0$ and solve $0 = 3x + 6$

$$3x = -6$$

$$x = -2$$

so the x-intercept is (–2, 0). Plot these points and draw your line. See Figure 2.11.

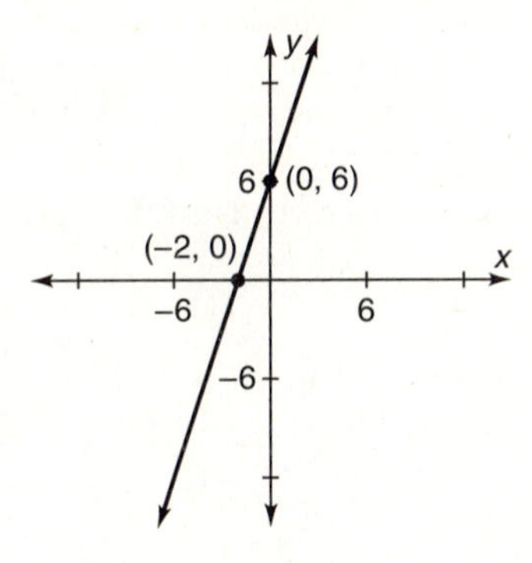

Figure 2.11

There may be a problem with graphing by using the two intercepts. There may not be two. Look at this next example.

EXAMPLE 2.7

Graph the line $y = 2x$.

When $x = 0$, then $y = 0$, and vice versa, so the x-intercept and the y-intercept are both the origin (0, 0). In this case, to get another number you can substitute your favorite number (as long as it's not zero) for x and find y. If −7 happens to be your favorite number, then

$$y = 2(-7) = -14$$

and your second point is (−7, −14). Plot and draw. See Figure 2.12.

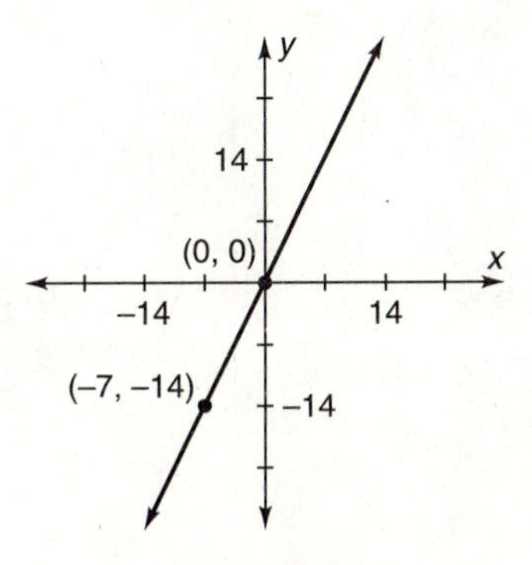

Figure 2.12

2.4 GRAPHING LINES WITH YOUR CALCULATOR

First, you must write your linear equation in the slope-intercept form

$$y = mx + b.$$

Then you can graph. We'll start with an easy example.

EXAMPLE 2.8

Graph $2x + 3y = 6$ using your calculator.

In slope-intercept form, it is

$$y = -\frac{2}{3}x + 3.$$

Enter the function in the calculator and press the **graph** key. You probably will see something like Figure 2.13. If your calculator graph doesn't look exactly like this picture, it's probably because your x and y ranges are set differently. In Figure 2.13, both x and y ranges are set from –10 to +10.

$$x \text{ min} = -10,$$

$$x \text{ max} = 10,$$

$$y \text{ min} = -10,$$

$$y \text{ max} = 10.$$

If you reset your ranges like this, then you will see exactly the same picture as Figure 2.13. To accomplish this press the **range** key, and enter these numbers.

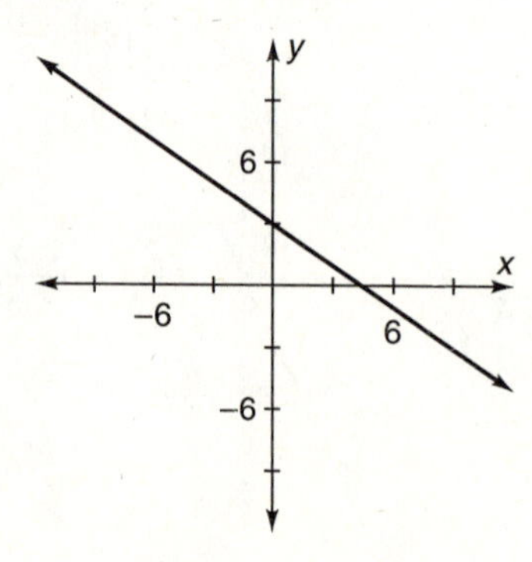

Figure 2.13

EXAMPLE 2.9

Graph $y = 10{,}000 - .02x$ on your calculator and show intercepts.

Enter the function and press **graph**. See anything other than axes? Your answer is "No!" unless you changed your ranges from the previous settings. Think about it: the y-intercept is 10,000, which is considerably larger than 10, and the x-intercept is 500,000, also bigger than 10. So this line cuts across the plane from (0, 10,000) to (500,000, 0), completely missing your screen! In order to see relevant information (intercepts) you should set $x \text{ min} < 0$, $x \text{ max} > 500{,}000$ and $y \text{ min} < 0$, $y \text{ max} > 10{,}000$. We choose

$$x \text{ min} = -100{,}000,$$

$$x \text{ max} = 600{,}000,$$

$$y \text{ min} = -500,$$

$$y \text{ max} = 12{,}000.$$

Here's the graph we get on our calculator with these ranges. See Figure 2.14. Note that these range settings are somewhat arbitrary and could be different—as long as relevant information is displayed (in this example, intercepts).

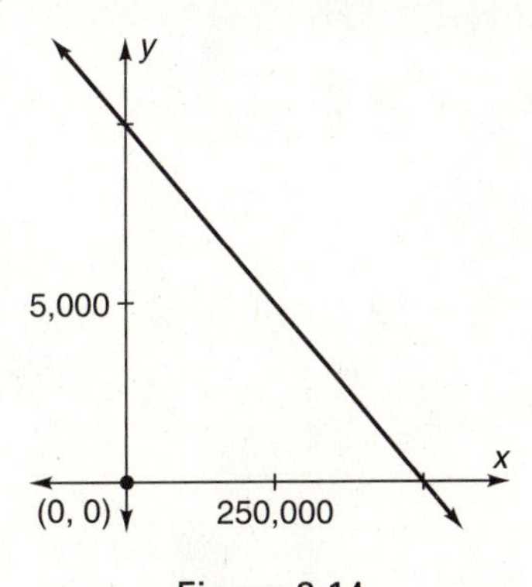

Figure 2.14

EXAMPLE 2.10

Suppose the function $P(t) = 2.374t - 4473.74$ defines the population P (in millions) of the United States in year $t \geq 1940$. For example, in 1950, the population was (substitute $t = 1950$)

$$P(1950) = 155.56 \text{ (in millions)},$$

or 155,560,000 people.

Graph this function on your calculator so that the graph shows the population for years 1940–2000.

Enter this function; if your calculator only graphs using variables x and y, you must use x for t, and y for P. Next, set x- and y-ranges. In this example, relevant information is contained between the years 1940 and 2000. So set the x-range to be

$$x \text{ min} \leq 1940,$$

$$x \text{ max} \geq 2000.$$

That was easy, but what about the y-range? First, note that the slope of this line is positive, so the function is increasing. This means the population for this time period was least in 1940, and will be greatest in 2000. In $t = 1940$,

$$P(1940) = 131.82$$

and in $t = 2000$,

$$P(2000) = 274.26.$$

Now set your y-range:

$$y \text{ min} = 130, y \text{ max} = 275,$$

and graph. You should see this picture. See Figure 2.15.

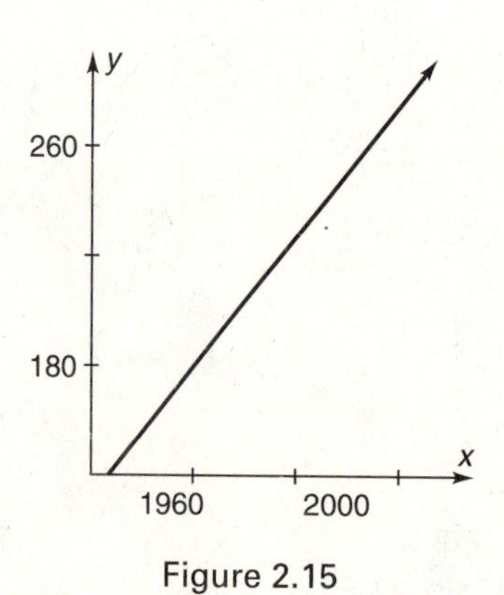

Figure 2.15

These calculator techniques are important, but you'll still need to know the slope and intercept for other reasons, so you'll need to learn them also.

2.5 WRITING YOUR OWN EQUATIONS

Until now, linear functions have appeared in front of you for you to graph, or analyze, or whatever. But the main thing you need to do for *Earth Algebra* is to make your own functions and equations. In order to write the equation of a line, you need certain information. In particular, either one of these two pieces of information will suffice:

1. two points on the line; or
2. one point on the line, and its slope.

Once you have one of these two things, you need to know what to do with it. This leads us to another type of equation for a line, so we'll discuss that.

Remember the slope formula

$$m = \frac{y_2 - y_1}{x_2 - x_1},$$

where (x_1, y_1) and (x_2, y_2) can be any two points on the line. Modify this slightly—hold (x_1, y_1) fixed (a known point), and let (x_2, y_2) be a variable point, and so as not to confuse it with a fixed point, leave off the subscripts and just call it (x, y). Now, multiply both sides of the equation $m = \frac{y - y_1}{x - x_1}$ by $x - x_1$, to get this wonderful equation

$$y - y_1 = m(x - x_1).$$

The reason this equation is so wonderful is that you can use it to write equations of lines that you need to study environmental problems. It is called the *point-slope equation for a line*, because if you look at it carefully, you can see a point on the line and the slope of the line. Those are, respectively, (x_1, y_1) and m.

If the equation,

$$y - 3 = 4(x - 5)$$

should appear before you, you look closely and see that (5, 3) is a point on the line, and that 4 is its slope!

If this equation fades from view and the new one,

$$y + 7 = 3(x - 1),$$

materializes, you know that (1, –7) is on this line, and the slope is 3.

Back to equation writing. Here's how to use the point-slope equation to write the equation of a particular line. Take the two cases, one by one.

EXAMPLE 2.11

Write the equation of the line which passes through (–1, 3) and has slope 5.
Substitute $x_1 = -1$, $y_1 = 3$ and $m = 5$ to get

$$y - 3 = 5(x - (-1)).$$

Simplify and solve for y, and the equation is

$$y = 5x + 8.$$

Easy!

Suppose you know two points. This time, you must do a tiny bit of work before substituting—you must compute the slope using the slope formula, then substitute either of your known points for (x_1, y_1) and your new-found slope m. Here's an example.

EXAMPLE 2.12

Suppose you know that the two points (2, –1) and (–4, 3) are on your unknown line. Write the equation of this line.
First, find m:

$$m = \frac{-1-3}{2-(-4)} = \frac{-4}{6} = -\frac{2}{3}.$$

Then use $x_1 = -4$, $y_1 = 3$ (or use the other point if you'd rather, you get the same answer) to get

$$y - 3 = -\frac{2}{3}(x - (-4)).$$

Simplify to

$$y - 3 = -\frac{2}{3}x - \frac{8}{3},$$

solve for y, and get

$$y = -\frac{2}{3}x + \frac{1}{3}.$$

If you prefer functional notation,

$$f(x) = -\frac{2}{3}x + \frac{1}{3}.$$

is your answer.

Below are practice problems for you to do.

2.6 THINGS TO DO WITH LINES AND LINEAR FUNCTIONS

For each of the Exercises 1–5, write the equation in slope-intercept form, and determine the slope and the y-intercept.

1. $y = -1.4x$.
2. $y - 5 = 3(x - 7)$.
3. $2(x - 4) = 3(y + 7)$.
4. $3x + 5y = 18$.
5. $1.2(3x + 5y) - 3.2(3x - 5) = 0$.

In Exercises 6–10, use the "three-second" technique to graph the line through the given point with the given slope.

6. $(2, -3)$, $m = 4$.
7. $(2, 5)$, $m = \dfrac{-3}{5}$.
8. $(1.7, 2.1)$, $m = -2$.
9. $(5, 12)$, $m = \dfrac{8}{3}$.
10. $(2.3, 5)$, $m = 0$.

Graph each of the functions below. Label the x-intercept and the y-intercept, and find the slope.

11. $f(x) = 2.4x + 12$.
12. $g(x) = 9 - .02x$.
13. $CO(t) = .0003t + 12$.
14. $r(t) = -102t + 17$.

15. $m(t) = 2(3t - 5) + .1(1 - 2t)$.

16. $T(x) = 1.8(x - 100) + 18$.

Write the equation for the line L satisfying the given information, then write in slope-intercept form. Graph the equation, and show both of the intercepts and determine the slope.

17. L passes through the points (10, 42.3) and (20, 51.5).

18. L passes through the point (0, –12) with slope 16.4.

19. L passes through the points (100, 2.2) and (150, 1.9).

20. L passes through the points (–30, 280) and (51, 350).

21. L has slope –.02, and passes through the point (125, 21.3).

22. L passes through the point (12, 2.4) with slope 0.

23. L passes through the points (–2, 5) and (2, –5).

24. L passes through the points (20, 18.5) and (48, 19.4).

25. L passes through the points (.0035, 2) and (.0012, 5).

CHAPTER 3

Atmospheric Carbon Dioxide Concentration

3.1 YOUR FIRST MODEL

There's more carbon dioxide than any other greenhouse gas. It is responsible for approximately fifty percent of the greenhouse effect. There's always been CO_2 in the atmosphere, and that's good—to a certain extent. But the concentration has been increasing significantly due to human activity. Here are just a few things that are most likely to be a part of your life which cause CO_2 to be emitted into the atmosphere: riding in your car to the mall (or anywhere); reading under the electric light; sitting in someone's teak chair; eating a hamburger at some fast-food joint; cranking up the furnace on a cold winter day, or the air conditioner on a hot summer day. The big sources of carbon dioxide are energy production and deforestation, both of which are for human consumption. We'll take a close look at some of these sources later, but now, for your first adventure in mathematical modeling, we look at atmospheric carbon dioxide concentration. In Chart 3.1 the data given is the CO_2 concentration in parts per million (ppm) by year. (Exercise: For the year 1980, put one million sky-blue dots on a piece of paper, then near the top, make three hundred thirty-eight and one half of them dirty and sooty.)

This data was compiled at the Mauna Loa Observatory in Hawaii.

Chart 3.1

Year	CO_2 Concentration (in ppm)
1965	319.9
1970	325.3
1980	338.5
1988	351.3

Now, here's a step-by-step procedure for modeling this data.

Step 1. Make points out of the data. The points have as first coordinate "year," and as second coordinate "CO_2 in ppm." In order to make the numbers more manageable, adjust the years in the first coordinate so that the year 1965 corresponds to 0; in other words, if t is the variable representing time, then $t = 0$ means 1965, and in general, t = number of years after 1965. Also, let CO_2C denote the variable for the second coordinate; i.e., CO_2C = carbon dioxide concentration in ppm. Functional notation would be this:

$$CO_2C(t) = \text{CO}_2 \text{ concentration in year } 1965 + t.$$

For example in 1988, CO_2 concentration was 351.3 ppm. This translates to

$$t = 23$$

and

$$CO_2C(23) = 351.3 \text{ ppm}.$$

The corresponding point is (23, 351.3).

Step 2. Set up a (t, CO_2C) coordinate system with t on the horizontal axis, CO_2C on the vertical axis, and plot the four points. (The zigzag at the base of the vertical axis indicates that part of the vertical axis has been left out.) It looks like Figure 3.1.

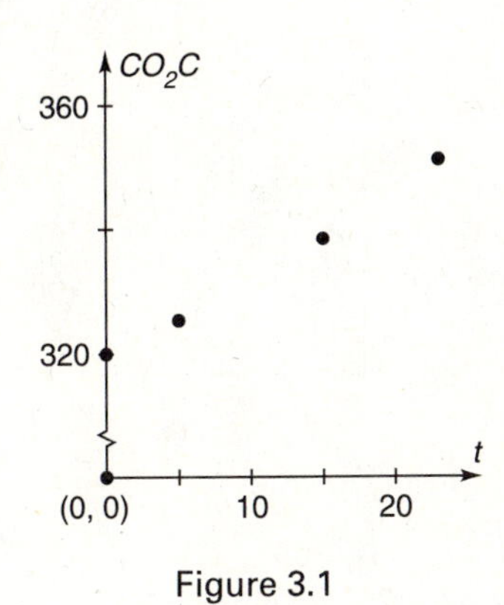

Figure 3.1

Step 3. In order to predict future concentrations of CO_2 based on past information, we need a function whose graph comes close to the points plotted. In this step, we simply "eyeball" these points to decide what kind of graph they are shaped like. These points appear to fall very close to a straight line. So, what we have to do is write the equation of a line which comes close to these points. There are many ways to do this, one of which is to choose two of the plotted points and write the equation of the line which passes through them. That's what we're going to do. First we try the years 1965 ($t = 0$) and 1988 ($t = 23$). The corresponding points are (0, 319.9) and (23, 351.3). The slope of the line through these points is

$$m = \frac{351.3 - 319.9}{23 - 0} = 1.37 \text{ (rounded)}$$

Now use the point-slope formula with point (0, 319.9):

$$CO_2C - 319.9 = 1.37\ (t - 0),$$

or

$$CO_2C = 1.37t + 319.9.$$

Using functional notation, we write

$$CO_2C(t) = 1.37t + 319.9.$$

Figure 3.2 is the graph of the original data and the "predictor" line whose equation we just wrote.

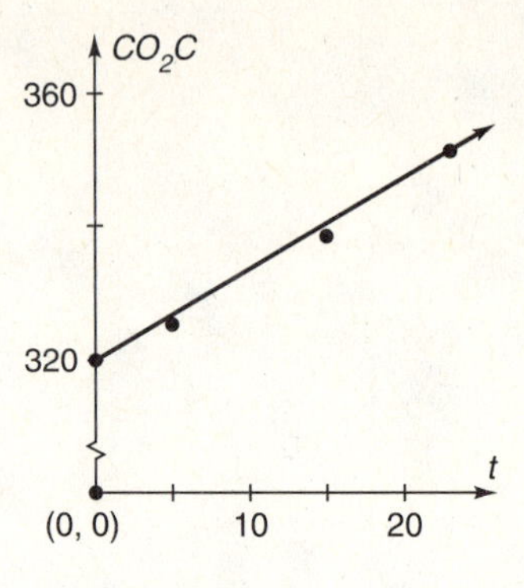

Figure 3.2

If we choose the two points (5, 325.3) and (15, 338.5) (corresponding to which years?), we get the linear function

$$CO_2C(t) = 1.32t + 318.7,$$

and here's its graph (Figure 3.3) along with the original data.

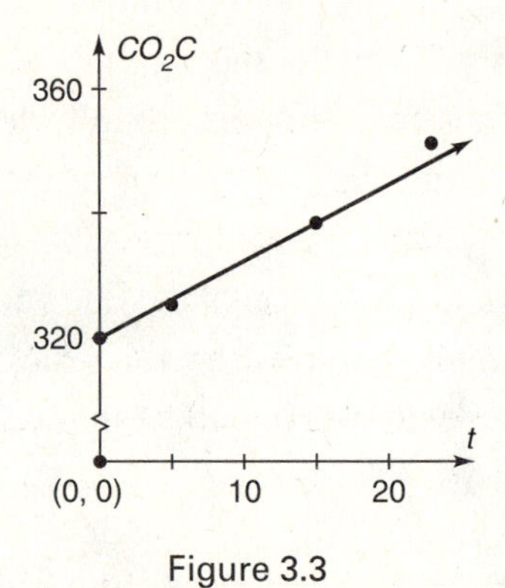

Figure 3.3

We're still on Step 3. There are a total of six linear equations to write by choosing pairs of points, and we've only done two. Be sure to use your calculator for the computations. You may even figure out a program which will do everything for you.

Derive the equations of the remaining lines.

Step 4. By the time we reach this step, we should have equations for all possible lines. We must now decide how to find the "best" line to use to model our data. Remember, we want the one which most closely approximates the real data. This means that the value of the derived function at each of the data years should be pretty close to the CO_2 concentration for that year. We check the first function,

$$CO_2C(t) = 1.37t + 319.9.$$

All evaluations in Chart 3.2 are done on the calculator.

Chart 3.2

Year	$CO_2C(t)$	Actual CO_2 Concentration
0	319.9	319.9
5	326.8	325.3
15	340.5	338.5
23	351.4	351.3

This first column is time t = years after 1965; the second column is the derived function evaluated at t (e.g., for year 1970, the "predicted" CO_2 concentration is $CO_2C(5) = 326.8$ ppm); and the third column is the original data. Next see how far off the "predicted" value is from the actual concentration. In Chart 3.3 below, "difference" refers to the distance between the predicted value and the actual value. Remember, the distance between two numbers is the absolute value of their difference.

Chart 3.3

Year	Difference
0	0.0
5	1.5
15	2.0
23	0.1

We used the data for 1965 and 1988 to derive the function

$$CO_2C(t) = 1.37t + 319.9,$$

so for $t = 0$ and 23, the difference should be zero, and it is for $t = 0$. Note, however, that it is 0.1 for $t = 23$; this is a result of round off. The differences for years which are used to derive the equation should always be small. (If these differences are large, you may want to check the derivation of your equation for mistakes.)

Next, we need to take into consideration the difference for each of the data years; we'd like all of these differences to be small for the best line, so we define an "overall" *error* for the line to be the sum of these differences. The error for our first derived predictor function is

$$E_1 = 0.0 + 1.5 + 2.0 + 0.1 = 3.6.$$

Of all possible predictor functions for the data provided, we define the best one to be the one with smallest error.

Here's the computation of the error for the second derived predictor function, $CO_2C(t) = 1.32t + 318.7$:

Chart 3.4

Year	$CO_2C(t)$	Actual CO_2 Concentration	Difference
0	318.7	319.9	1.2
5	325.3	325.3	0.0
15	338.5	338.5	0.0
23	349.1	351.3	2.2

Add the column on the right; if the answer is smaller than $E_1 = 3.6$, then Function 2 is the winner. $E_2 = 3.4$, so Function 2 is better than Function 1.

In summary, the four steps to modeling are:

1. make points out of the data;
2. set up a coordinate system and plot the points;
3. decide on the type of predictor curve to use by observing the shape of the

plotted points. Then derive all possible equations. (This is the long step, but without your calculator, you might spend all weekend doing it.)

4. Compute the error for each predictor function to determine the best function. This is also kind of long, but the calculator makes it easy. Be sure to clearly state what the variables mean.

There are four more lines, so we may not have the best one yet. Compute the errors for all possible predictor functions and determine the best one.

Obviously, carbon dioxide concentration in the atmosphere is related to CO_2 emission from human activity. Therefore, a model of population would be interesting. The chart below gives U.S. population in millions in the indicated year. Model this with a linear equation using $t = 0$ in 1960. You should go through the four steps listed in this section.

Year	U.S. Population ($\times 10^6$)
1960	179.3
1970	203.3
1980	226.5
1987	243.4
1990	250.0

3.2 USING YOUR FIRST MODEL

First, we list (Chart 3.5) all of the linear equations derived from the CO_2 concentration data, together with the error for each. You can use this listing to check your own derivations, and your best equation.

Chart 3.5

Years	Points	Equation: $CO_2C(t)$ =	Error
1965, 1970	(0, 319.9), (5, 325)	$1.08t + 319.9$	9.0
1965, 1980	(0, 319.9), (15, 338.5)	$1.24t + 319.9$	3.7
1965, 1988	(0, 319.9), (23, 351.3)	$1.37t + 319.9$	3.6
1970, 1980	(5, 325.3), (15, 338.5)	$1.32t + 318.7$	3.4
1970, 1988	(5, 325.3), (23, 351.3)	$1.44t + 318.1$	3.1
1980, 1988	(15, 338.5), (23, 351.3)	$1.60t + 314.5$	8.2

The smallest error occurs for years 1970 and 1988, and hence the best function for this model is

$$CO_2C(t) = 1.44t + 318.1.$$

Remember, $CO_2C(t)$ = carbon dioxide concentration in ppm in year $1965 + t$. It is interesting to interpret the slope of this equation: each year the CO_2 concentration will increase by 1.44 ppm.

Here's how this model can be used to predict future atmospheric CO_2 levels.

EXAMPLE 3.1

In the year 2000, $t = 2000 - 1965 = 35$, and the predicted CO_2 concentration is $CO_2C(35) = 368.5$ ppm.

EXAMPLE 3.2

In the year 2050, $t = 2050 - 1965 = 85$, and the predicted CO_2 concentration is $CO_2C(85) = 440.5$ ppm.

EXAMPLE 3.3

The preindustrial carbon dioxide level of concentration was 280 ppm; double this amount is significant, as you will see in the next chapter. Your model can be used to predict when this doubling will occur. Twice 280 is 560 ppm, so we should determine t when $CO_2C(t) = 560$, or solve this equation:

$$560 = 1.44t + 318.1$$
$$241.9 = 1.44t$$
$$t = \frac{241.9}{1.44}$$
$$t = 168 \text{ years}$$

This is the year 1965 + 168, or 2133.

The preindustrial carbon dioxide concentration level of 280 ppm actually remained relatively constant until World War II. Let's see how closely our model reflects this information; i.e., according to our equation, in what year was $CO_2C = 280$? Thus, we must solve the equation

$$280 = 1.44t + 318.1$$
$$-38.1 = 1.44t$$
$$t = \frac{-38.1}{1.44}$$
$$t = -26.5 \text{ years}$$

This corresponds to mid 1938 (1965 – 26.5 = 1938.5). Not a bad approximation, since the war actually started in 1939!

Notice that this model for CO_2 concentration yields carbon dioxide levels which are below the actual level of 280 ppm for all years before 1938; hence the equation is valid only after 1938. For this reason, we adjust this equation so that $t = 0$ in 1939, and then restrict its domain to $t \geq 0$. This adjustment can be performed by replacing t in $1.44t + 318.1$ by $t - 26.5$ (recall $t = -26.5$ when $CO_2C = 280$). We get

$$CO_2C\,(t - 26.5) = 1.44\,(t - 26.5) + 318.1$$
$$= 1.44t + 280.$$

We still call this last expression $CO_2C(t)$, but now

$$CO_2C(t) = 1.44t + 280,$$

where $CO_2C(\text{t})$ = carbon dioxide concentration in ppm in year

$1939 + t,\ t \geq 0.$

This is the model we use for the remainder of Part I.

Use this equation to answer the following questions.

1. Predict CO_2 concentration in the year 2000.
2. In what year will the concentration be double the pre-industrial level? (For Questions 1 and 2 , you should have gotten the same answers as the ones in the last section using the old equation.)
3. Predict CO_2 concentration in year 2061. (This is the year Halley's Comet returns.)
4. In what year will CO_2 concentration be 420 ppm?
5. Discuss what social, political or physical changes might affect the accuracy of this model.

Use your U.S. population model (see data at end of Section 3.1) to answer the following questions.

1. Predict the U.S. population in 2020.
2. When will the U.S. population double its present population?
3. Write a verbal interpretation of the slope of your population equation, i.e., how much does the U.S. population increase annually?
4. Predict the U.S. population in the year you were born. You might want to go to the library and check the accuracy of this prediction in the Statistical Abstracts.
5. Currently, the U.S. population is 4.7% of world population. Assuming that this percentage remains constant, predict the year in which world population will reach 8 billion.
6. Discuss any social, political or other factors which night affect the accuracy of your model.

In the next chapters, you'll see when some of the bad stuff starts, i.e., some consequences of more and more CO_2 in the atmosphere.

CHAPTER 4

Composite Functions

4.1 DEFINITION AND EXAMPLES

Finding a composite function is as easy as evaluating a function. Actually, it's almost the same as the operation of evaluation of a function at a number, except you substitute another function for x instead of a number. Just remember the cat ( ): whatever goes in the parentheses for x also replaces x in the mathematical expression.

If $f(x)$ and $g(x)$ are two functions, then $f(g(x))$ denotes the composite function obtained by substituting the function $g(x)$ in for the x in the function $f(x)$. This is read "f of g of x." You may be wondering why anyone would ever need to find a composite. Often it is desirable to express mathematically a relation between two entities. Remember the earlier function $F(w)$ which defined the number F of catfish in a North Georgia lake in terms of the number w of gallons of toxic waste dumped upstream. Well, there's also a restaurant on that lake which serves fresh catfish, and the price of a dinner depends on the availability of fish from the lake. So think about the function $P(F)$ which defines the price P of a catfish dinner in terms of the number of fish in the lake. (Note: this would be a decreasing function surely; the fewer the fish, the higher the price.) The real cause of a price increase is the amount of toxic waste dumped. So, the function

Note: Chapter 4 is a prerequisite for Chapter 5.

which would describe this is the composite $P(F(w))$; this defines the price of a catfish dinner in terms of the amount of toxic waste dumped upstream! Time for an example.

EXAMPLE 4.1

Let $f(x) = x^2 + 4x + 3$, and $g(x) = 2x$. To find $f(g(x))$ substitute $g(x)$ (which equals $2x$) for x in $f(x)$ to get

$$f(g(x)) = f(2x) = (2x)^2 + 4(2x) + 3 = 4x^2 + 8x + 3.$$

Also, $g(f(x))$ is a composite. This time you substitute $f(x)$ (which equals $x^2 + 4x + 3$) for x in $g(x)$ to get

$$g(f(x)) = g(x^2 + 4x + 3) = 2(x^2 + 4x + 3) = 2x^2 + 8x + 6.$$

One more example, and then you can practice with some of the exercises below.

EXAMPLE 4.2

Let $f(x) = 3x - 1$ and $g(x) = x^2 + 7$.

$$f(g(x)) = f(x^2 + 7) = 3(x^2 + 7) - 1 = 3x^2 + 20$$

and

$$g(f(x)) = g(3x - 1) = (3x - 1)^2 + 7 = 9x^2 - 6x + 1 + 7 = 9x^2 - 6x + 8.$$

One thing you should note: a composite of two functions yields another function. There is no equation to solve or anything like that. (Students for thousands and thousands of years have found a composite, then set it equal to zero and tried to solve for x; who knows why?)

4.2 THINGS TO DO

Determine the composites.

1. $f(x) = 3x - 5$, $g(x) = 2 - x$. Determine $f(g(x))$ and $g(f(x))$.
2. $f(x) = 5(2 - 3x)$; $H(x) = 3 + (2x - 5)$. Determine $H(f(x))$, $f(H(x))$, $f(H(0))$, and $H(f(.2))$.
3. $r(x) = \frac{1}{x}$; $s(x) = 5x + 2$. Determine $r(s(x))$ and $s(r(x))$.

4. $AB(t) = t + 2$; $C(s) = s^2 - 2$. Determine $C(AB(-2))$, $AB(C(-2))$, $C(AB(t))$ and $AB(C(s))$.

5. $E(x) = x$; $H(x) = .13578x^3 - .00023$. Determine $E(H(x))$, $H(E(x))$, $E(H(0))$, $E(H(1))$, $H(E(1))$.

6. $PQ(x) = 3x + 1.5$; $QP(x) = \dfrac{x - 1.5}{3}$. Determine $PQ(QP(x))$ and $QP(PQ(x))$.

7. A chemical factory has set up business on a certain North Georgia river. This river empties into a large lake, on which is a restaurant which serves fresh catfish dinners. The chemical factory started dumping toxic waste into the river, which reduced the catfish population. This is described by the equation

$$F(w) = 50 - .5w,$$

where F = number of fish per acre and w = gallons of toxic waste dumped per day. In turn, the reduced availability of catfish caused the restaurant to increase the price of its dinner. This is described by

$$P(F) = 24 - .4F, \text{ where } P = \text{price of a catfish dinner.}$$

   a. Determine the composite function which defines the price of a catfish dinner in terms of the number w of gallons of toxic waste dumped by the chemical factory.

   b. How much did a dinner cost before the factory opened?

   c. How much waste will correspond to a price of \$12?

   d. How much waste will kill all the fish?

CHAPTER 5

Global Temperature and Ocean Level

5.1 ON THE BEACH...

If the global temperature rises, then water expands, and polar ice caps would melt, thus causing ocean levels to rise. Should this happen, our beaches will disappear.

It is estimated that a doubling of the pre-industrial level of atmospheric CO_2 concentration will cause an average global temperature increase of 5.4° F, and furthermore, an increase of as little as 1.8° F in global temperature can cause a one foot rise in ocean levels. That's pretty significant!

What we are going to do now is to study what will actually happen to the beaches if CO_2 emission continues to increase at its current rate. First, we write an equation which relates temperature increase to atmospheric CO_2 concentration. We derive the equation using the estimate that doubling the pre-industrial level of CO_2 causes a 5.4° F temperature increase, and assume the relationship is linear. We use a (CO_2C, GT) coordinate system, where the variables

$$CO_2C = \text{carbon dioxide concentration in ppm},$$

and

$$GT = \text{global temperature increase since 1939};$$

i.e., $GT = 0°$ F in 1939. (Remember the significance of the year 1939 from

Chapter 3?) To write our equation, we need two points. The first one comes from the preindustrial level CO_2 concentration, which was 280 ppm, and the corresponding global temperature increase is 0° F; so the point is (280, 0). To get the second point, use the information above: double 280 ppm is 560 ppm, and the corresponding $GT = 5.4°$ F, so this point is (560, 5.4). The slope of the line is

$$m = \frac{5.4 - 0}{560 - 280} = .0193,$$

and the equation is

$$GT - 0 = .0193(CO_2C - 280)$$

or

$$GT = .0193CO_2C - 5.4.$$

In functional notation,

$$GT(CO_2C) = .0193CO_2C - 5.4,$$

where

$$GT(CO_2C) = \text{global temperature increase},$$

$$CO_2C = \text{carbon dioxide concentration in ppm.}$$

The significance of the slope of the equation is this: an increase in CO_2 concentration of one ppm corresponds to an increase in global temperature of .0193° F.

Next, we can relate global temperature increase to year by using our $CO_2C(t)$ model derived in Chapter 3:

$$CO_2C(t) = 1.44t + 280,$$

where

$$CO_2C(t) = \text{carbon dioxide concentration in ppm in year } 1939 + t.$$

We define this relationship between CO_2C and t by finding the composite function $GT[CO_2C(t)]$.

$$GT[CO_2C(t)] = .0193\ (1.44t + 280) - 5.4$$

or

$$GT[CO_2C(t)] = .028t.$$

Since this defines global temperature in terms of year, we suppress CO_2C from the notation and write

$$GT(t) = .028t,$$

where $GT(t)$ = average global temperature increase in year $1939 + t$. Notice that the slope of this linear function tells you that the average global temperature will increase .028° F each year.

Write a linear equation which defines ocean level increase OL in terms of global temperature increase GT; this defines a function $OL(GT)$. To derive this equation, use the information that a temperature increase of 1.8° F corresponds to a one-foot rise in ocean level. Also, assume that in 1939 temperature increase is 0° F, so that oceans have not risen any. This will give you two points.

After deriving $OL(GT)$, form the composite $OL(GT(CO_2C))$ which defines ocean level increase in terms of carbon dioxide concentration; this will be $OL(CO_2C)$.

Next, obtain the important equation $OL(t)$, which defines ocean level increase in terms of time t. To derive this function, find your $CO_2C(t)$ function where $t = 0$ in 1939, and form the composite $OL(CO_2C(t))$ which gives $OL(t)$. From this linear equation $OL(t)$, determine how much the oceans will rise each year.

This last equation, $OL(t)$, is quite valuable. It gives us a tool for figuring how many years it will take for the ocean to rise a given distance. For example, if it is important to know when the ocean will have risen three and one-half feet above its 1939 level, then simply take $OL(t) = 3.5$, solve the equation for t, and add the answer to 1939.

5.2 OR, WHAT BEACH?

On to the beach! In 1991, the authors drove to Tybee Island (near Savannah, Georgia) and estimated the average height of the sea wall on the beach there to be four feet, and the base of this wall to be 2.2 feet above the ocean level. See the picture on the next page.

Interesting Fact. Did you know that tall buildings affect weather patterns? Some years ago, thunderstorms and rain clouds which developed over the Gulf of Mexico would move inland over the west coast of Florida, proceed across the state and out over the Atlantic providing ample rainfall for inland areas and the eastern coast. However, in the past twenty years or so, there has been extensive development of high rise hotels and condominiums along the eastern shore. These large buildings create thermals and cause weather systems to turn north as they come inland so that they never reach the eastern and central areas of Florida. The result has been periodic drought conditions in some portions of the state. In particular, the Everglades have suffered considerably from these conditions, as well as from other human development projects such as damming and irrigation methods upstate.

Find out what year the ocean will completely cover the beach at Tybee Island. Next, it would also be interesting to know when the ocean will be over the sea wall and flood the development. When will this happen?

PART II

Factors Contributing to Carbon Dioxide Build-up

INTRODUCTION

In Part II you will study three major sources of carbon dioxide emission: automobiles in the United States, energy consumption in the United States, and deforestation.

If you live in the suburbs near a big city, and work a nine-to-five job in the city, then you probably already know about carbon dioxide emission from automobiles. If you do not drive to work and home every day in that sea of metal, through that fog of carbon compounds, then may you never have to.

These cars are one of the major sources of carbon dioxide in the United States. The number of automobiles continues to increase sharply, and Americans are driving more and more each year. Although most manufacturers are improving the fuel efficiency of their vehicles, the resulting improvements are not fast enough to balance CO_2 emissions. It was estimated that in the year 1991, over 750 million tons of carbon dioxide were pumped into the atmosphere from passenger vehicles in this country. This figure represents an 18% increase over the corresponding emission ten years ago. Don't breathe too deeply.

The United States is responsible for 25% of the global emission of carbon dioxide, and 20% of that is due to automobiles. Thus, automobiles in the U.S. contribute 5% of the world's CO_2 emission.

In Chapter 8 we study the output of carbon dioxide from automobiles in the United States. Three sets of data are easily available from the *Statistical Abstract of the United States*: number of cars by year; the average number of miles each car travels per year; and the average gas mileage per car by year. (Copies of the

Statistical Abstract are kept in most libraries, and are updated each year.) These three pieces of information, together with the fact that each gallon of gasoline burned results in 20 pounds of CO_2 in the atmosphere, can be combined to determine the annual carbon dioxide emission.

The gasoline you burn in your car is just one source of carbon dioxide emission. Every time you turn on a light bulb or run your furnace, you use energy. It takes energy to manufacture the paper that this is written on and to grow the food you eat. The production and consumption of energy is one of the major causes of greenhouse gas emissions. In the United States, the annual average consumption of energy is equivalent to the consumption of 2200 gallons of petroleum per person. For comparison, the average person in Japan uses the equivalent of about 1000 gallons of petroleum per person, and in the developing countries such as Brazil, China, or India, the consumption is the equivalent of 150 gallons per capita. Energy consumption dropped in the early 1980s but since then it has been on the rise.

In Chapter 9 energy consumption data in the United States are provided (*Statistical Abstract of the United States*) for coal, petroleum, and natural gas. Each set of data is to be modeled by the groups, and finally total energy consumption will be determined.

Another major factor affecting global warming is the alarming rate of destruction of tropical rainforests. Almost all of this destruction, or deforestation, is done for human consumption.

Many acres of rainforest are being cleared and burned for homesites and farming by locals, and then more land needs to be cleared to build access roads, and on and on. It took years upon hundreds of years for rainforests to form, and once destroyed they don't grow back. When the trees are cut and burned, they release tons of carbon into the atmosphere, thus adding to the already increasing volume of greenhouse gases. Trees also absorb carbon when growing, so when forests are cut, much of the carbon which would have been absorbed remains in the atmosphere. Tropical forests can absorb over 80 metric tons of carbon per hectare per year, and when cut release up to one-half metric ton of carbon for each hectare. Therefore, deforestation has two negative effects on a stable planetary climate.

And besides all this, rainforests are very beautiful. And so are the animals that make their homes there....

In Chapter 11 we categorize most of the direct reasons for deforestation into three big areas: (1) logging; (2) cattle grazing; and (3) agriculture, mining, and development. Indirectly, these factors are related to population growth and increasing demand for forest products. We analyze each of these areas to find out its effect on the overall loss of rainforests and on atmospheric carbon dioxide.

In Chapter 12, we build models which define total emission of carbon from each of the three sources, automobiles, energy consumption, and deforestation.

The mathematical concepts used to conduct these studies in Part II are linear, quadratic, logarithmic and piecewise functions, systems of equations, and matrices. Also, exponential functions are introduced for use later.

CHAPTER 6

Quadratic Functions

6.1 NOTATION AND DEFINITIONS

You are moving up a notch! In the preceding section, you studied linear functions. The largest exponent used for a linear function is 1. Quadratic functions have x raised to the second power: x^2. A *quadratic function* looks like

$$f(x) = ax^2 + bx + c,$$

where a, b, and c are just constants, but important ones. The equation above is referred to as the *standard form* for a quadratic function. The graph of a quadratic function is called a *parabola* (pa rab' o la, not pa ra bo' la), and it looks like one of the two pictures in Figure 6.1.

You get the picture on the left when the leading coefficient a is positive, and people say that this parabola *opens up*. On the other hand, if $a < 0$, the parabola on the right occurs, and then people say that it *opens down*.

Every parabola in the universe has a *vertex*, which is quite important. The vertex is the lowest point ($a > 0$), or the highest point ($a < 0$) on the graph of a quadratic function. See Figure 6.2.

If you draw a vertical line through the vertex of a parabola, its graph on one side of the line is the mirror image of its graph on the other side; i.e., it is symmetric about that vertical line.

Note: Chapter 6 is a prerequisite for Chapters 8 and 9.

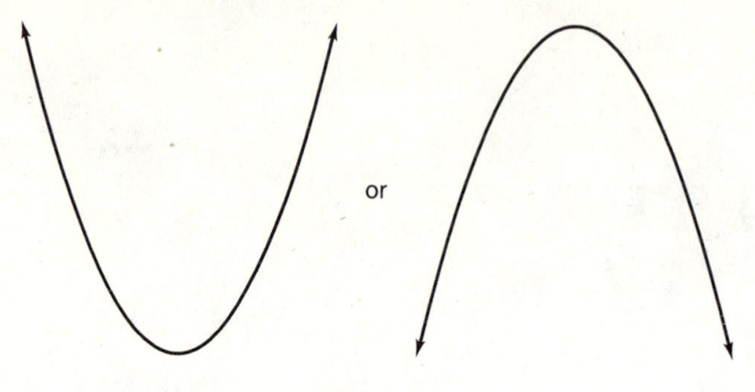

Figure 6.1

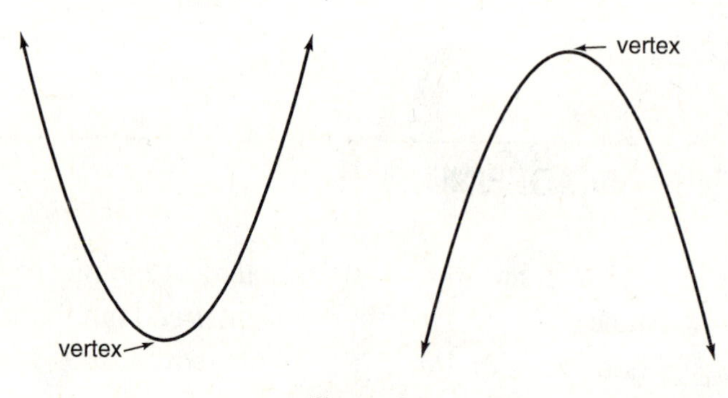

Figure 6.2

Other points that are really important on a parabola are its intercepts. A parabola can have as many as three intercepts, one on the y-axis, and possibly two on the x-axis (recall that intercepts are points where a graph crosses an axis).

The domain of a quadratic function consists of all real numbers x, and the range, well, we'll discuss that later when we talk about the vertex in more detail.

Right now, we concentrate on the important features of the graph of a quadratic function, $y = ax^2 + bx + c$:

1. open up or down?

2. y-intercept

3. x-intercept(s)

4. vertex

and

5. maximum or minimum value (more on this below).

The important thing about those constants a, b, and c is that you can draw the graph and provide those important five items by just doing some arithmetic with a, b, and c, or letting your calculator do it for you. We'll show you how, step-by-step.

Step 1. Open up or down? This is simple. Put your function in standard form, and look at a. If $a > 0$, it opens up. If $a < 0$, it opens down.

In the following sections, we go through the remaining items with you, so you can be an expert parabola artist.

6.2 INTERCEPTS

Step 2. y-intercept. Easy. Substitute zero for x in the equation $y = ax^2 + bx + c$ to get the answer. Remember, you're looking for the place where the graph crosses the y-axis, which is where $x = 0$.

Step 3. x-intercept, or x-intercepts, or, are there any? Now you're looking for the places where the graph intersects the x-axis, and this happens where $y = 0$. So, you substitute $y = 0$, and then you have to solve the quadratic equation

$$0 = ax^2 + bx + c.$$

There are three possibilities which can occur. We look at each graphically for a parabola which opens up. See Figure 6.3.

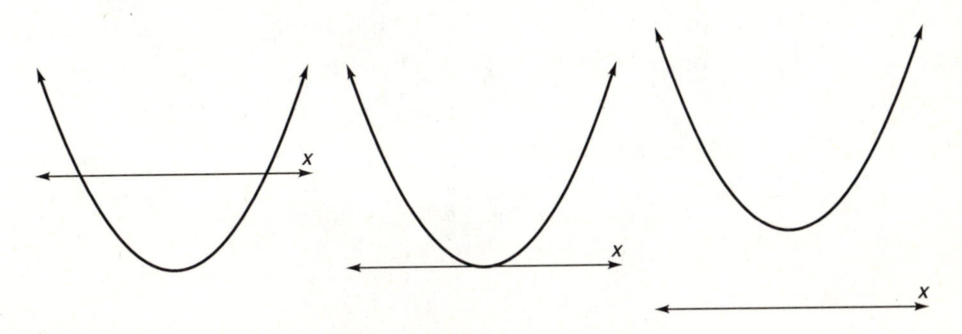

Figure 6.3

Of course the same three situations can happen for a parabola which opens down—turn the page over and upside down, hold it up against a window and look at these graphs.

When you get two x-intercepts, there will be two solutions to

$$0 = ax^2 + bx + c;$$

when there is one x-intercept, there'll be one solution; and when there is no x-intercept, the equation has no solution.

We review ways to solve quadratic equations. Probably factoring is what comes immediately to your mind. For example: solve $0 = x^2 - x - 2$. Factor $0 = (x - 2)(x + 1)$, and set each factor equal to zero.

$$x - 2 = 0, \quad x + 1 = 0;$$

$$x = 2 \text{ and } x = -1.$$

However, there's not one single quadratic equation in *Earth Algebra* that even comes remotely close to factoring.

But, remember the quadratic formula? (See Appendix C for its derivation.) This can be used to solve any quadratic equation, if it has a solution. It always works. *Always.* Just put your equation in standard form, pick off the coefficients a, b, and c and plug them into

$$x = \frac{-b \pm \sqrt{b^2 - 4ac}}{2a}.$$

This is the celebrated *quadratic formula*; if you don't know it, or have forgotten it, learn it now. You'll need it throughout *Earth Algebra.*

There's one thing you must remember always to do before solving a quadratic equation—no matter how it presents itself, you must put it in standard form before using the quadratic formula. Here are some examples.

EXAMPLE 6.1

Find the x-intercepts of the quadratic function $y = -9.6 - 3.2x + 1.5x^2$.

Set $y = 0$,

$$0 = -9.6 - 3.2x + 1.5x^2,$$

and solve.

Use the quadratic formula, but first rearrange to get

$$0 = 1.5x^2 - 3.2x - 9.6.$$

Now $a = 1.5$, $b = -3.2$, $c = -9.6$, and

$$x = \frac{-(-3.2) \pm \sqrt{(-3.2)^2 - 4(1.5)(-9.6)}}{2(1.5)}$$

Get out your calculator! Your answer is (rounded to two places) $x = 3.81$ and $x = -1.68$. Graphically, the quadratic $f(x) = 1.5x^2 - 3.2x - 9.6$ has two x-intercepts. See Figure 6.4.

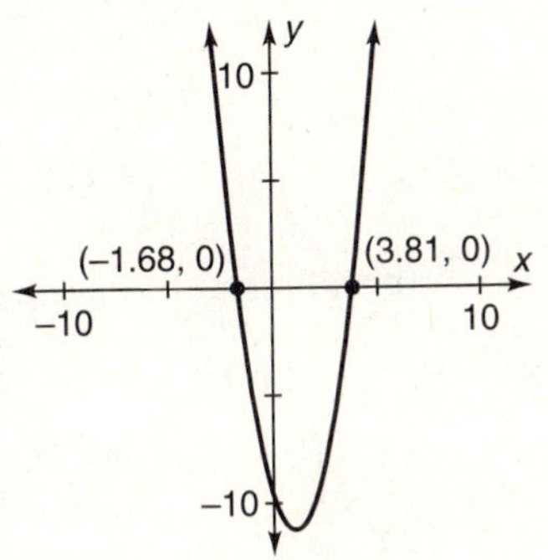

Figure 6.4

EXAMPLE 6.2

Determine the x-intercepts of the quadratic function

$$f(x) = -2.3x^2 + 1.1x - 27.6.$$

Replace $f(x)$ by 0 and solve the quadratic equation

$$0 = -2.3x^2 + 1.1x - 27.6.$$

The equation is already in standard form, and $a = -2.3$, $b = 1.1$, $c = -27.6$.

$$x = \frac{-1.1 \pm \sqrt{(1.1)^2 - 4(-2.3)(-27.6)}}{2(-2.3)}$$

And, the calculator does your arithmetic and tells you that you have erred! Your error is in trying to take the square root of -252.71. No one can take the square root of a negative number and get a real number for an answer. This means that your original quadratic equation has no real solution. (In *Earth Algebra*, the only numbers we use are real.) Graphically (Figure 6.5), this happens when there are no x-intercepts.

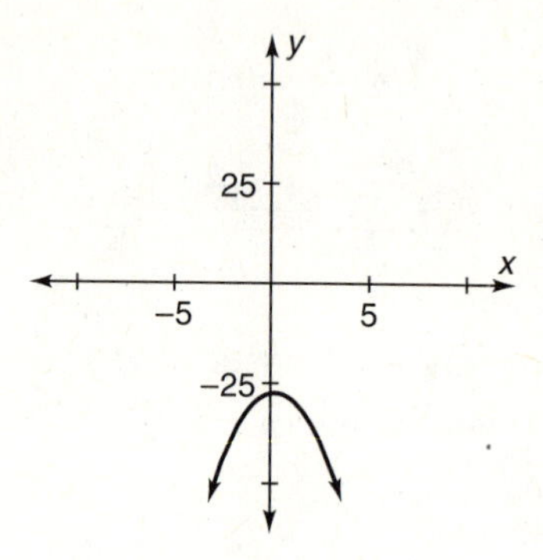

Figure 6.5

EXAMPLE 6.3

Determine the x-intercepts for this one:

$$y = (x^2 - 9.2x + 26.01) - x.$$

Standard form first:

$$0 = x^2 - 10.2x + 26.01.$$

Now $a = 1$, $b = -10.2$, $c = 26.01$, and

$$x = \frac{-(10.2) \pm \sqrt{(-10.2)^2 - 4(1)(26.01)}}{2(1)}$$

The calculator tells you that the quantity under the radical is zero, so the $\pm$ doesn't matter. There's only one answer: $x = 5.1$. Graphically, this means that the parabola has (5.1, 0) as its only x-intercept. See Figure 6.6.

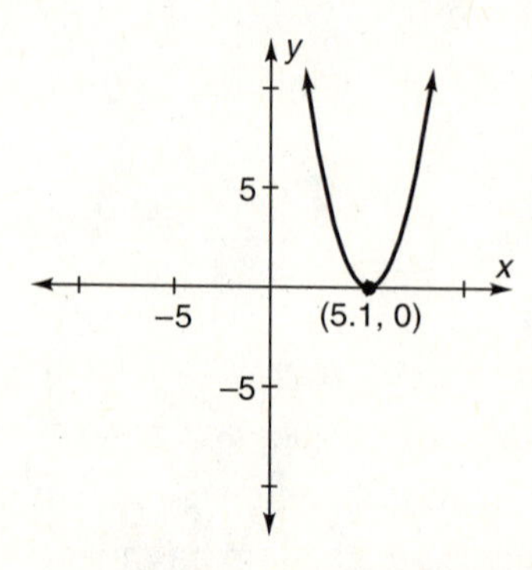

Figure 6.6

After completing this section, turn to Exercises 1–8 in Section 6.7. Decide whether each of these parabolas opens upward or downward and determine the coordinates of the x- and y-intercepts of each.

6.3 VERTEX

Step 4. Determine the coordinates of the vertex. The first two coefficients, a and b, tell you what the first coordinate of the vertex is. The formula is:

$$x = \frac{-b}{2a}.$$

You get the second, or y-coordinate by substituting $\frac{-b}{2a}$ into the function; i.e., $y = f\left(\frac{-b}{2a}\right)$. The coordinates of the vertex are $\left(\frac{-b}{2a}, f\left(\frac{-b}{2a}\right)\right)$. Compute $\frac{-b}{2a}$ on the calculator, then evaluate the quadratic function at this number.

EXAMPLE 6.4

Find the vertex of the parabola $f(x) = -2.6x^2 + 7.6x - 10$.

The x-coordinate of the vertex is 1.46 (rounded), and $y = f(1.46) = -4.45$ (also rounded) is its y-coordinate. The vertex is the point $(1.46, -4.45)$. In this case, it is the highest point on the graph. Note that the graph opens down because $a = -2.6 < 0$. See Figure 6.7. (If $a > 0$, the parabola opens up, and the vertex will be the lowest point on the graph.)

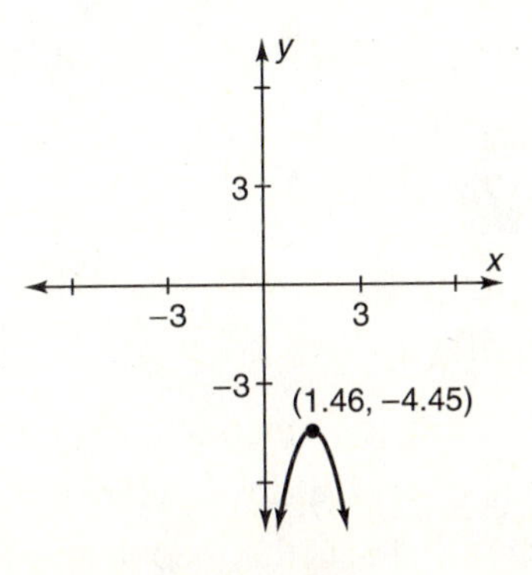

Figure 6.7

Now turn again to Exercises 1–8 (Section 6.7) and determine the vertex of each parabola.

6.4 MAXIMUM, MINIMUM, AND RANGE OF A QUADRATIC FUNCTION

Step 5. Find the maximum or minimum value. The y-coordinate of any point on a graph tells how high (or low) the point is. Since the second coordinate of a point on the graph of a function is $f(x)$, then the highest point on the graph has the largest possible y-coordinate, or functional value, that can occur. So a quadratic function has a maximum when its graph opens down $(a < 0)$, and that maximum is the y-coordinate of its vertex. Similarly, if a parabola opens up, it has a minimum which is the y-coordinate of its vertex.

For Example 6.4 above, $f(x) = -2.6x^2 + 7.6x - 10$ has a maximum value of -4.45, which occurs when $x = 1.46$. See Figure 6.7.

Maximum and minimum values of functions are really important throughout *Earth Algebra*. If you know the maximum or the minimum of a function, you can often determine its range. The function in Example 6.4,

$$f(x) = -2.6x^2 + 7.6x - 10,$$

has as its maximum value -4.45, and has no minimum value, so its range consists of all $y \leq -4.45$.

EXAMPLE 6.5

Find the vertex, maximum or minimum, and range of

$$f(x) = 1.2x^2 + 3.7x - 5.9.$$

The x-coordinate of the vertex is

$$x = \frac{-b}{2a} = \frac{-3.7}{2(1.2)} = -1.54,$$

and its y-coordinate is

$$y = f(-1.54) = -8.75.$$

To decide whether the function has a maximum or a minimum, we need to know if the graph opens up or down. Here $a = 1.2 > 0$, so it opens up and the function has a minimum: this is the y-coordinate of the vertex, or -8.75.

The function has no maximum, so its range consists of all $y \geq -8.75$. See Figure 6.8.

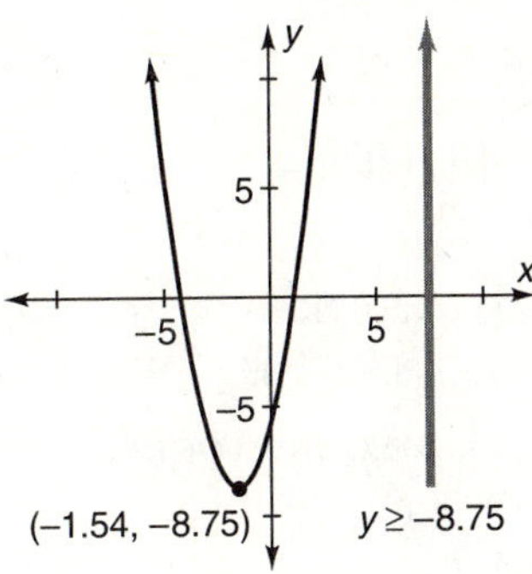

Figure 6.8

If a number k is in the range of a quadratic function, we can solve equations of the form $f(x) = k$. In Example 6.5 above, we can solve the equation $f(x) = 1.4$ since $1.4 \geq -8.75$. Replace $f(x)$ by 1.4 and solve

$$1.4 = 1.2x^2 + 3.7x - 5.9.$$

First put the equation in standard form

$$0 = 1.2x^2 + 3.7x - 7.3,$$

and use the quadratic formula to solve. Now $a = 1.2$, $b = 3.7$ and $c = -7.3$, giving us solutions $x = 1.37$ and $x = -4.45$. This tells you that you have two solutions to the equation

$$1.4 = 1.2x^2 + 3.7x - 5.9,$$

$x = 1.37$ and $x = -4.45$. See Figure 6.9.

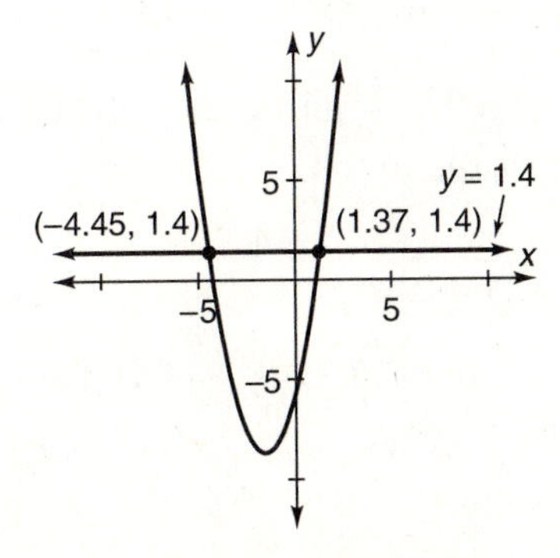

Figure 6.9

Now turn to Exercises 1–8 in Section 6.7 and determine the maximum or minimum value and range of each of these quadratic functions.

6.5 GRAPHS OF QUADRATIC FUNCTIONS

In the next two examples, 6.6 and 6.7, we graph quadratic functions from start to finish, and provide all significant information. This information consists of vertex, intercepts, maximum or minimum, and range.

EXAMPLE 6.6

Let $f(x) = -1.1x^2 + 30.2x - 197$.

First, $a = -1.1 < 0$, so the graph opens down. Its y-intercept is –197, which occurs when $x = 0$. To find its x-intercepts, solve the equation,

$$0 = -1.1x^2 + 30.2x - 197,$$

by using the quadratic formula.

$$x = \frac{-30.2 \pm \sqrt{(30.2)^2 - 4(-1.1)(-197)}}{2(-1.1)},$$

$$x = 10.67 \text{ and } x = 16.78.$$

Next, we find its vertex. The x-coordinate is

$$x = \frac{-30.2}{2(-1.1)} = 13.73,$$

and the y-coordinate is

$$f(13.73) = 10.28.$$

This function has a maximum of 10.28, and its range consists of all $y \leq 10.28$. Figure 6.10 shows the graph.

EXAMPLE 6.7

Graph $f(x) = 0.4x^2 - x + 7.3$, and solve the equation $f(x) = 5$.

This one opens up because $a = 0.4 > 0$. Its y-intercept is 7.3, and x-intercepts are obtained as solutions to

$$0 = 0.4x^2 - x + 7.3.$$

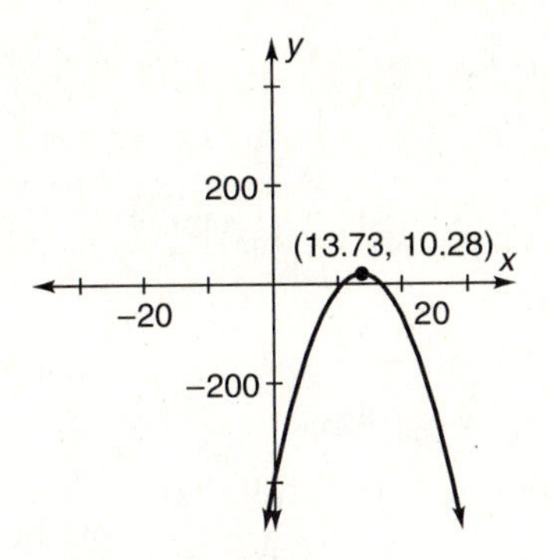

Figure 6.10

Substitution into the quadratic formula gives

$$x = \frac{-(1.1) \pm \sqrt{(-1)^2 - 4(0.4)(7.3)}}{2(0.4)}.$$

The quantity under the radical is −10.68, a negative number. Hence this equation has no solution; i.e., the graph has no x-intercepts. (This does not mean there is no graph, only that it does not cross the x-axis.)

The vertex has coordinates

$$x = \frac{-(-1)}{2(0.4)} = 1.25,$$

and $y = f(1.25) = 6.675$. (Helpful hint: if we had first found the vertex, we would have known that there are no x-intercepts. The y-coordinate of the vertex is positive, and the graph opens up, so it is impossible for y to be zero anywhere on this graph!) This function has a minimum of 6.675, and its range consists of all numbers $y \geq 6.675$. This tells us that the equation $f(x) = 5$ has no solution. See Figure 6.11.

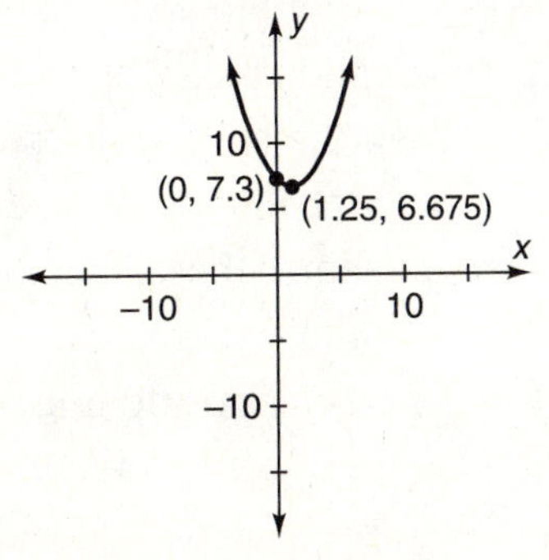

Figure 6.11

Now turn to Section 6.7 and graphs 1–8 and answer Questions 9–12.

6.6 YOU AND YOUR CALCULATOR GRAPHS

Your calculator will graph just about any function. Also, it will estimate the coordinates of any point on the graph. But, before you begin your graph, you need to know certain information about the function. We provide you with illustrative examples.

EXAMPLE 6.8

Graph $f(x) = 0.9x^2 + 8.2x - 239.7$ on your calculator. Be sure that vertex and intercepts are visible. Also, find vertex coordinates and estimate x-intercepts with the calculator.

Enter the function and press **graph**. You probably don't see anything except axes. This is because you are looking at the wrong part of the xy-plane. To find a suitable area to see the important graph points, you need to do some work. First, calculate the vertex:

$$x = \frac{-8.2}{2(0.9)} = -4.56 \ (\textit{rounded}),$$

and

$$y = f(-4.56) = -258.38 \text{ (evaluate with calculator).}$$

This parabola opens up, so the entire graph will lie above $y = -258.38$, and there must be two x-intercepts. Set the y-range to y min $= -300$ (or smaller than -258.38) and y max $= 50$ (or any positive number that allows viewing the graph above the x-axis); then return to your graph. Next press the **trace** key and move the cursor along the graph to the right until it crosses the x-axis; set x max to be any number larger than the calculator estimate of this intercept (x max $= 20$). Now move the cursor to the left to estimate the smaller x-intercept, and set x min to be any number smaller than the calculator estimate, say x min $= -25$. Return to your graph, and you should see Figure 6.12. Finally, you can estimate the coordinates of the x-intercepts by positioning the cursor as close as possible to one intercept, and using the "zoom" feature of your calculator. The number $x = 12.39$ is a good estimate for the larger one, and $x = -21.50$ is a good estimate of the smaller.

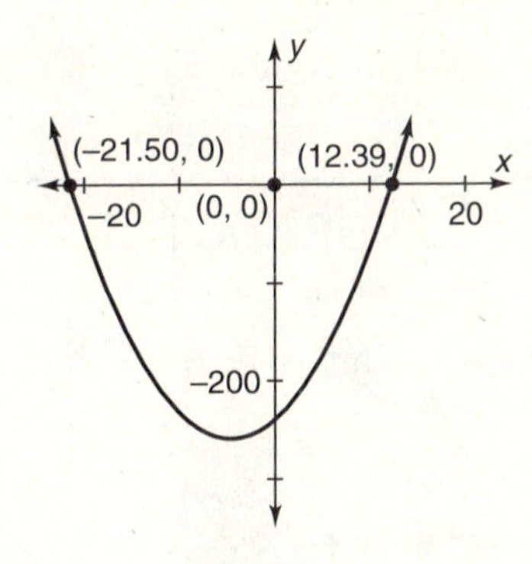

Figure 6.12

Actually, you can estimate the coordinates of any point on a graph using the "zoom" feature. For example, see how close you can estimate the vertex of this parabola using this technique.

EXAMPLE 6.9

Graph $f(x) = -0.03x^2 + .11x - 1.2$ on your calculator so that vertex and intercepts are visible. Also estimate coordinates of these important points.

Enter this function and press **graph**. Unless your x range is very small, you should see the vertex on your screen. The graph opens down, and the vertex is below the x-axis, so there are no x-intercepts. You can use your calculator to estimate the coordinates of the vertex. With the **trace** key, position the cursor on the graph as close as possible to the vertex, and use the "zoom" feature. You will know when the cursor is near the vertex if the y-coordinates shown on the screen increase, then decrease as the cursor moves from left to right. See Figure 6.13. A good estimate is (1.87, –1.099).

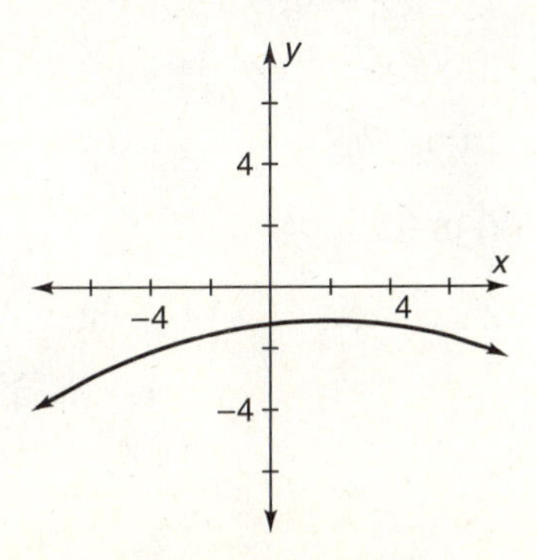

Figure 6.13

Do Exercises 13–16 in Section 6.7 after completing this section.

6.7 THINGS TO DO WITH QUADRATIC FUNCTIONS

These exercises are to be done as you work your way through this chapter.

1. $AIR(x) = x^2 - 6x + 5$.
2. $BP(x) = 5x^2 + 6x + 15$.
3. $CAR(x) = .02\,x^2 - 63.1x + 27$.
4. $DF(t) = 3200\,t^2 - 450t + 12$.
5. $E(z) = -3z^2 + 27.5z + 13.1$.
6. $F(x) = .0025x^2 - 6x + 3600$.
7. $GB(x) = 9 - 4x^2 + .6x$.
8. $IR(x) = .0027x^2 + 1.2x - .13$.

Answer Exercises 9–12; the functions are the ones defined in Exercises 1–8.

9. When is $BP(x) = 100$? (Ref. #2.)
10. When is $DF(t) = -10$? (Ref. #4.)
11. When is $E(z) = 50$? (Ref. #5.)
12. When is $F(x) = 60$? (Ref. #6.)

Graph the quadratic functions in Exercises 13–16 using the calculator. Use the **trace** key on the calculator to arrive at good estimates for the vertex and any other points you are asked to find.

13. $HD(x) = .0006x^2 - .126x + 12.9$.
14. $Joe(y) = -3.2(y^2 - 1.1y + 4) + 5$.
15. $K(b) = .003b^2 - 1.09b + 21.2$; when is $K(b) = 45$? 50?
16. $L(t) = -12.2t^2 + 54.6t - 1.6$; when is $L(t) = 30$? 40? 50? 60?

CHAPTER 7

Systems of Linear Equations and Matrices

7.1 SYSTEMS OF LINEAR EQUATIONS

A linear equation in n variables can be written in the form

$$a_1x_1 + a_2x_2 + \ldots + a_nx_n = c,$$

where $x_1, x_2, \ldots, x_n$ are the variables and $a_1, a_2, \ldots, a_n, c$ are constants. A system of m linear equations in n variables can be written in this form:

$$\begin{aligned} a_{11}x_1 + a_{12}x_2 + \ldots + a_{1n}x_n &= c_1 \\ a_{21}x_1 + a_{22}x_2 + \ldots + a_{2n}x_n &= c_2 \\ &\cdot \\ &\cdot \\ &\cdot \\ a_{m1}x_1 + a_{m2}x_2 + \ldots + a_{mn}x_n &= c_m, \end{aligned}$$

where, once again, $x_1, \ldots, x_n$ are the variables, and a_{ij}, $i = 1, 2, \ldots, m, j = 1, 2, \ldots, n$, and c_k, $k = 1, \ldots, m$, are constants. This is called an $m \times n$ system. Notice that the

---

Note: Chapter 7 is a prerequisite for Chapters 8 and 9.

constant a_{ij} is the coefficient of the jth variable x_j in the ith equation; for example, a_{34} is the coefficient of x_4 in the third equation. These subscripts will be important later.

A solution to a system of linear equations is a set of n numbers

$$x_1 = b_1, x_2 = b_2, \ldots, x_n = b_n$$

which make all the equations true.

We focus on methods of finding solutions to 2×2 and 3×3 systems. The system

$$\begin{aligned} x + y &= 5 \\ 3x - 4y &= 1 \end{aligned}$$

is a 2×2 system, whereas

$$\begin{aligned} x + 2y + z &= 0 \\ 2x - y - 3z &= 1 \\ 3x - 2y \quad &= -1 \end{aligned}$$

is a 3×3 system.

When solving a system of linear equations, the following three operations are used.

1. Interchange two equations.

2. Replace any equation by a non-zero multiple of that equation.

3. Replace any equation by itself plus a multiple of another.

Solving a system of equations using these operations is called solving by elimination of variables.

EXAMPLE 7.1

Solve the 2×2 system

$$\begin{aligned} x + y &= 5 \\ 3x - 4y &= 1. \end{aligned}$$

The variable y can be eliminated by replacing the second equation by itself plus 4 times the first equation:

$$\begin{aligned} &4(x + y = 5) \\ &+ (3x - 4y = 1), \end{aligned}$$

or

$$\begin{aligned} &4x + 4y = 20 \\ &+ (3x - 4y = 1), \end{aligned}$$

so the new second equation is

$$7x = 21,$$

and hence

$$x = 3.$$

To find y, substitute $x = 3$ into either of the original equations; if we choose the first, we get

$$3 + y = 5,$$

so

$$y = 2.$$

The solution to this system is $x = 3$, $y = 2$.

Geometrically, this solution gives the coordinates of the point of intersection of the graph of the two equations.

EXAMPLE 7.2

Solve the system

$$\begin{aligned} 3x - y &= 4 \\ -6x + 2y &= 10. \end{aligned}$$

We eliminate x in the second equation by multiplying the first equation by 2 and adding to the second to get

$$0 = 18.$$

Everyone knows that 0 can't equal 18. This means that this system has no solution. Geometrically the lines which are the graphs of these two equations are parallel.

EXAMPLE 7.3

Solve the 3×3 system

$$x - y + z = 8$$
$$x + 2y - z = -5$$
$$2x + 3y + z = 6.$$

There's more to do here because there are more variables. First we eliminate z from the second equation by adding the first and second to get

$$2x + y = 3.$$

To get another equation in x and y, eliminate z from the third equation by adding the third and (–1) times the first; this gives

$$x + 4y = -2.$$

Now, we can solve the resulting 2×2 system,

$$2x + y = 3$$
$$x + 4y = -2.$$

Multiply the second equation by –2 and add to the first to get

$$-7y = 7,$$

so

$$y = -1.$$

Substitute this $y = -1$ into $x + 4y = -2$ (or into $2x + y = 3$) to get

$$x - 4 = -2,$$

or

$$x = 2.$$

Finally substitute $x = 2$ and $y = -1$ into any one of the original three equations to find z. If we choose the first, we get

$$2 - (-1) + z = 8,$$

so

$$z = 5.$$

Therefore, the solution is

$$x = 2, y = -1, z = 5.$$

EXAMPLE 7.4

Solve the system

$$x + 2y + z = 4$$
$$x + 3y - 5z = -2$$
$$2x - y - 2z = 4.$$

We eliminate x from the second and third equation to get a 2×2 system in y and z. First, subtract the first from the second equation to get

$$y - 6z = -6.$$

Next subtract twice the first from the third to get

$$-5y - 4z = -4.$$

Now solve the 2×2 system

$$y - 6z = -6$$
$$-5y - 4z = -4$$

to get

$$y = 0, z = 1.$$

Substitute these variables into the first original equation (or any other one) to see that

$$x = 3.$$

Turn to Section 7.4 and solve Exercises 1–10.

7.2 MATRICES

This is what you need to know about matrices: a matrix is a rectangular array of numbers. Its size is designated by the number of rows and columns it has; for example a matrix with four rows and six columns is a 4×6 matrix. The numbers in the matrix are called entries; the number in the ith row and j^{th} column is the i, j-entry; for example, the number in the second row and third column is the 2,3-entry. The general form for an $m \times n$ matrix is

$$\begin{bmatrix} a_{11} & a_{12} & \cdots & a_{1n} \\ a_{21} & a_{22} & \cdots & a_{2n} \\ \cdot & \cdot & \cdot & \cdot \\ a_{m1} & a_{m2} & a_{m3} & a_{mn} \end{bmatrix},$$

where a_{ij} is the i, j-entry.

An example of a 3×4 matrix is

$$\begin{bmatrix} -1 & -6 & 8 & -2 \\ 0 & 7 & 14 & 10 \\ 2 & -3 & 0 & -3 \end{bmatrix}$$

Its 2,3-entry is $a_{23} = 14$; its 3,1-entry is $a_{31} = 2$, etc., etc.

Matrices can be added together and subtracted from each other as long as they are the same size: add or subtract corresponding entries. We give an example of each of these operations.

EXAMPLE 7.5

$$\begin{bmatrix} 1 & 0 & 4 \\ -2 & 6 & -1 \end{bmatrix} + \begin{bmatrix} 5 & 1 & -9 \\ 10 & 2 & 3 \end{bmatrix} = \begin{bmatrix} 6 & 1 & -5 \\ 8 & 4 & 2 \end{bmatrix}.$$

EXAMPLE 7.6

$$\begin{bmatrix} 3 & 1 & 2 \\ 0 & -1 & 4 \\ 7 & 0 & 8 \end{bmatrix} - \begin{bmatrix} 5 & 2 & -3 \\ 4 & 0 & 6 \\ 1 & -1 & 2 \end{bmatrix} = \begin{bmatrix} -2 & -1 & 5 \\ -4 & -1 & -2 \\ 6 & 1 & 6 \end{bmatrix}.$$

A matrix can also be multiplied by any real number; multiply each entry by the real number. This operation is called scalar multiplication.

EXAMPLE 7.7

$$3\begin{bmatrix} -1 & 0 \\ 7 & -4 \\ 2 & 8 \end{bmatrix} = \begin{bmatrix} -3 & 0 \\ 21 & -12 \\ 6 & 24 \end{bmatrix}.$$

Another very important operation on matrices is matrix multiplication. If two matrices are the right size, then they can be mulitplied together to produce a new matrix. If A and B are matrices, in order to perform the product AB, the matrix A must have the same number of columns as the number of rows of the matrix B. For the purpose of illustration, this first example is simple.

EXAMPLE 7.8

$$[1 \quad 2 \quad 3]\begin{bmatrix} 4 \\ 5 \\ 6 \end{bmatrix} = [1\cdot 4 + 2\cdot 5 + 3\cdot 6] = [32].$$

Here $A = [1 \quad 2 \quad 3]$ is 1×3, and $B = \begin{bmatrix} 4 \\ 5 \\ 6 \end{bmatrix}$ is 3×1, and their product $AB = [32]$ is 1×1.

In general, if A is $m \times k$ and B is $k \times n$, then the product AB is $m \times n$. This product matrix is defined by multiplying the ith row of A by the jth column of B (multiply corresponding entries, and then sum) to produce the i,j-entry of AB (as in example 7.8).

EXAMPLE 7.9

$$\begin{bmatrix} 1 & -2 \\ 4 & -3 \end{bmatrix}\begin{bmatrix} 4 & -1 & 0 \\ 0 & 2 & 5 \end{bmatrix} = \begin{bmatrix} 1\cdot 4 + (-2)0 & 1(-1) + (-2)2 & 1\cdot 0 + (-2)5 \\ 4\cdot 4 + (-3)0 & 4(-1) + (-3)2 & 4\cdot 0 + (-3)5 \end{bmatrix}$$

$$= \begin{bmatrix} 4 & -5 & -10 \\ 16 & -10 & -15 \end{bmatrix}.$$

EXAMPLE 7.10

$$\begin{bmatrix} 1 & 2 & 3 \\ 4 & 5 & 6 \\ 7 & 8 & 9 \end{bmatrix} \cdot \begin{bmatrix} 1 & 0 & 0 \\ 0 & 1 & 0 \\ 0 & 0 & 1 \end{bmatrix} = \begin{bmatrix} 1 & 2 & 3 \\ 4 & 5 & 6 \\ 7 & 8 & 9 \end{bmatrix},$$

and

$$\begin{bmatrix} 1 & 0 & 0 \\ 0 & 1 & 0 \\ 0 & 0 & 1 \end{bmatrix} \cdot \begin{bmatrix} 1 & 2 & 3 \\ 4 & 5 & 6 \\ 7 & 8 & 9 \end{bmatrix} = \begin{bmatrix} 1 & 2 & 3 \\ 4 & 5 & 6 \\ 7 & 8 & 9 \end{bmatrix}.$$

EXAMPLE 7.11

$$\begin{bmatrix} 2 & 5 \\ 1 & 3 \end{bmatrix} \cdot \begin{bmatrix} 3 & -5 \\ -1 & 2 \end{bmatrix} = \begin{bmatrix} 1 & 0 \\ 0 & 1 \end{bmatrix},$$

and

$$\begin{bmatrix} 3 & -5 \\ -1 & 2 \end{bmatrix} \cdot \begin{bmatrix} 2 & 5 \\ 1 & 3 \end{bmatrix} = \begin{bmatrix} 1 & 0 \\ 0 & 1 \end{bmatrix}.$$

In example 7.10 the matrix with all the 0s and 1s is special. It is called the identity matrix. The $n \times n$ identity matrix has 1s on its diagonal, and 0s elsewhere. If A and B are matrices and I is an identity matrix such that the products AI and IB are defined, then $AI = A$ and $IB = B$.

In example 7.11, the matrix

$$A = \begin{bmatrix} 2 & 5 \\ 1 & 3 \end{bmatrix}$$

has what is known as an inverse matrix. It is

$$A^{-1} = \begin{bmatrix} 3 & -5 \\ -1 & 2 \end{bmatrix}.$$

From this same example, you see that

$$AA^{-1} = \mathrm{I}\ (2 \times 2 \text{ identity}),$$

and also that

$$A^{-1}A = I.$$

In general, an $n \times n$ matrix has an inverse if there is an $n \times n$ matrix $B = A^{-1}$ such that

$$AB = BA = I\ (n \times n \text{ identity}).$$

Only square matrices ($n \times n$) have inverses, and not all of these do. Determining the inverse of a matrix (if it has one) is beyond the scope of this book.

Now turn to Section 7.4 and solve Exercises 11–32.

7.3 SYSTEMS OF EQUATIONS AND MATRICES

In this section, we illustrate the relationship between matrices and systems of equations, and how a system of linear equations can sometimes be solved using matrix methods.

Consider the system of 3 equations in 3 variables,

$$4x + y - 2z = 7$$

$$-3x - y + 2z = 3$$

$$5x + y - 3z = 4.$$

We make a matrix of coefficents for the unknowns x, y, and z. There will be one row for each equation, and one column for each unknown. The first row contains coefficients of the first equation; for example, the 1,1-entry is 4, which is the coefficient of the first unknown x in the first equation. The 1,2-entry is 1, which is the coefficient of the second unknown y in the first equation. To complete the first row, the 1,3-entry is –2. Here's the matrix, called the coefficient matrix of the system:

$$A = \begin{bmatrix} 4 & 1 & -2 \\ -3 & -1 & 2 \\ 5 & 1 & -3 \end{bmatrix}.$$

This is a 3×3 matrix. Its 3,1-entry is 5, and 5 is the coefficient of x in the third equation.

The other numbers in this system of equations are the constants on the right hand side of the equal marks; put them in a 3×1 matrix, which is called the matrix of constants for the system:

$$B = \begin{bmatrix} 7 \\ 3 \\ 4 \end{bmatrix}.$$

Notice that there is one row for each equation and the entry in each row is the constant in the corresponding equation. Next, put the unknowns in another 3×1 matrix

$$X = \begin{bmatrix} x \\ y \\ z \end{bmatrix}.$$

The product

$$AX = B$$

produces the original system of equations:

$$\begin{bmatrix} 4 & 1 & -2 \\ -3 & -1 & 2 \\ 5 & 1 & -3 \end{bmatrix} \begin{bmatrix} x \\ y \\ z \end{bmatrix} = \begin{bmatrix} 4x + y - 2z \\ -3x - y + 2z \\ 5x + y - 3z \end{bmatrix} = \begin{bmatrix} 7 \\ 3 \\ 4 \end{bmatrix}.$$

The coefficient matrix A in this example does have an inverse

$$A^{-1} = \begin{bmatrix} 1 & 1 & 0 \\ 1 & -2 & -2 \\ 2 & 1 & -1 \end{bmatrix}.$$

If we multiply both sides of the matrix equation $AX = B$ by A^{-1}, here's what happens:

$$\begin{bmatrix} 1 & 1 & 0 \\ 1 & -2 & -2 \\ 2 & 1 & 1 \end{bmatrix} \begin{bmatrix} 4 & 1 & -2 \\ -3 & -1 & 2 \\ 5 & 1 & -3 \end{bmatrix} \begin{bmatrix} x \\ y \\ z \end{bmatrix} = \begin{bmatrix} 1 & 1 & 0 \\ 1 & -2 & -2 \\ 2 & 1 & -1 \end{bmatrix} \begin{bmatrix} 7 \\ 3 \\ 4 \end{bmatrix}$$

$$\begin{bmatrix} 1 & 0 & 0 \\ 0 & 1 & 0 \\ 0 & 0 & 1 \end{bmatrix} \begin{bmatrix} x \\ y \\ z \end{bmatrix} = \begin{bmatrix} 10 \\ -7 \\ 13 \end{bmatrix}$$

so

$$\begin{bmatrix} x \\ y \\ z \end{bmatrix} = \begin{bmatrix} 10 \\ -7 \\ 13 \end{bmatrix},$$

which means the solution to the original system is

$$x = 10, y = -7, z = 13.$$

What you have just seen always works as long as the coefficient matrix A is square ($n \times n$) and has an inverse. That is, if A is the coefficient matrix of a system of n equations in n variables with inverse A^{-1}, and if x is the column matrix consisting of the n variables, and B is the column matrix consisting of the n constants, then the original system is represented by the matrix equation

$$AX = B,$$

and the solution matrix is determined by

$$X = A^{-1}B.$$

EXAMPLE 7.12

Solve the system of equations

$$3x + 3y - z = 11$$

$$-2x - 2y + z = 4$$

$$-4x - 5y + 2z = 2$$

using matrix methods.

The coefficient matrix is

$$A = \begin{bmatrix} 3 & 3 & -1 \\ -2 & -2 & 1 \\ -4 & -5 & 2 \end{bmatrix};$$

the matrix of variables is

$$X = \begin{bmatrix} x \\ y \\ z \end{bmatrix};$$

and the matrix of constants is

$$B = \begin{bmatrix} 11 \\ 4 \\ 2 \end{bmatrix}.$$

We give you the inverse of A:

$$A^{-1} = \begin{bmatrix} 1 & -1 & 1 \\ 0 & 2 & -1 \\ 2 & 3 & 0 \end{bmatrix}.$$

Now, the solution is determined by the product

$$X = A^{-1}B = \begin{bmatrix} 1 & -1 & 1 \\ 0 & 2 & -1 \\ 2 & 3 & 0 \end{bmatrix} \begin{bmatrix} 11 \\ 4 \\ 2 \end{bmatrix} = \begin{bmatrix} 9 \\ 6 \\ 34 \end{bmatrix},$$

so

$$x = 9, y = 6, z = 34.$$

Your calculator will solve a system of equations for you. Here's how: first, set the dimensions of A and B. You do not need to enter X; the product $A^{-1}B$ will be X. Next, enter the matrices A and B, and finally, perform the product $A^{-1}B$ to see the solution matrix. (This method only works if A has an inverse.) There might be even easier ways to solve systems of equations on your calculator. Read the manual.

EXAMPLE 7.13

Solve the system of equations

$$x + 5y - z = 3$$

$$4x + 3y - 2z = 5$$

$$2x - 2y + z = 9$$

using your calculator.

The coefficient matrix is 3×3,

$$A = \begin{bmatrix} 1 & 5 & -1 \\ 4 & 3 & -2 \\ 2 & -2 & 1 \end{bmatrix};$$

the constant matrix is 3×1,

$$B = \begin{bmatrix} 3 \\ 5 \\ 9 \end{bmatrix}.$$

Set dimensions of each; enter each matrix, and take the product to get

$$A^{-1}B = \begin{bmatrix} 3 \\ 1 \\ 5 \end{bmatrix},$$

so $x = 3$, $y = 1$, and $z = 5$.

EXAMPLE 7.14

Solve the system using your calculator,

$$4x + y = 7$$

$$2x + 3y = -1.$$

The coefficeint matrix is 2×2,

$$A = \begin{bmatrix} 4 & 1 \\ 2 & 3 \end{bmatrix};$$

the constant matrix is 2×1,

$$B = \begin{bmatrix} 7 \\ -1 \end{bmatrix}.$$

The product matrix is

$$A^{-1}B = \begin{bmatrix} 2.2 \\ -1.8 \end{bmatrix},$$

so

$$x = 2.2, \; y = -1.8.$$

7.4 THINGS TO DO

Solve the following systems of equations by elimination of variables.

1. $\begin{aligned} x + 3y &= 1 \\ 2x - 3y &= 2 \end{aligned}$

2. $\begin{aligned} 4x + y &= 5 \\ 8x + 3y &= 2 \end{aligned}$

3. $\begin{aligned} x - y &= 1 \\ 2x + 3y &= -2 \end{aligned}$

4. $\begin{aligned} x - y &= 3 \\ 2x + 3y &= 1 \end{aligned}$

5. $\begin{aligned} 4x + y &= 1 \\ 2x + 3y &= -2 \end{aligned}$

6. $\begin{aligned} x + y &= 6 \\ y + z &= 10 \\ x + z &= 18 \end{aligned}$

7. $\begin{aligned} x + 2y + z &= 3 \\ x + 3y + 4z &= 6 \\ 2x + 4y + 3z &= 7 \end{aligned}$

8. $\begin{aligned} 2x + y + z &= 7 \\ 2x + y - z &= 5 \\ 2x + 4y + 3z &= 12 \end{aligned}$

9. $\begin{aligned} 2x + y + 3z &= 1 \\ 3x - y - 2z &= 3 \\ 4x - 3y - 7z &= 6 \end{aligned}$

10. $\begin{aligned} 2x + 3y - z &= 0 \\ 3x - 2y + z &= 7 \\ 4x + 6y - 3z &= 5 \end{aligned}$

In Exercises 11–29, perform the indicated operation whenever it is defined.

11. $\begin{bmatrix} 3 & 1 \\ 2 & 6 \\ 1 & 3 \end{bmatrix} + \begin{bmatrix} 0 & 2 \\ 1 & -1 \\ 7 & 5 \end{bmatrix}$

12. $\begin{bmatrix} 3 & 1 \\ 2 & 6 \\ 1 & 3 \end{bmatrix} + \begin{bmatrix} 0 & 2 \\ 1 & -1 \\ 7 & 5 \end{bmatrix}$

13. $\begin{bmatrix} 1 & 3 \\ 2 & 4 \end{bmatrix} - \begin{bmatrix} 4 & 3 \\ 2 & 1 \end{bmatrix}$

14. $\begin{bmatrix} 1 & 0 \\ 0 & 1 \end{bmatrix} + \begin{bmatrix} 0 & 0 & 0 \\ 0 & 0 & 0 \end{bmatrix}$

15. $5\begin{bmatrix} 2 & 1 \\ 3 & 6 \end{bmatrix}$

16. $-2\begin{bmatrix} 1 & 1 & 1 \\ 3 & 2 & 5 \end{bmatrix}$

17. $2\begin{bmatrix} 3 & 1 \\ 2 & 0 \end{bmatrix} - 4\begin{bmatrix} 0 & 2 \\ 1 & 3 \end{bmatrix}$

18. $\begin{bmatrix} 0 & 0 \end{bmatrix} + \begin{bmatrix} 3 & 7 \end{bmatrix}$

19. $\begin{bmatrix} 1 & 3 & 2 \end{bmatrix} + \begin{bmatrix} 2 \\ 3 \\ 7 \end{bmatrix}$

20. $\begin{bmatrix} 3 & 1 \end{bmatrix}\begin{bmatrix} 2 \\ 5 \end{bmatrix}$

21. $\begin{bmatrix} 1 & 2 \\ 3 & 1 \end{bmatrix}\begin{bmatrix} 5 \\ 2 \end{bmatrix}$

22. $\begin{bmatrix} 2 \\ 5 \end{bmatrix}\begin{bmatrix} 3 & 1 \end{bmatrix}$

23. $\begin{bmatrix} 3 & 2 & 4 \end{bmatrix}\begin{bmatrix} 1 \\ 3 \end{bmatrix}$

24. $\begin{bmatrix} 1 & 6 & 2 \\ 6 & 1 & 4 \\ -1 & 2 & 1 \end{bmatrix}\begin{bmatrix} 3 \\ 2 \\ -6 \end{bmatrix}$

25. $\begin{bmatrix} 2 & 3 \\ 1 & 5 \end{bmatrix}\begin{bmatrix} 2 \\ 1 \end{bmatrix}$

26. $\begin{bmatrix} 23 & 29 \\ 17 & 43 \end{bmatrix}\begin{bmatrix} 1 & 0 \\ 0 & 1 \end{bmatrix}$

27. $\begin{bmatrix} 1 & 0 & 0 \\ 0 & 1 & 0 \\ 0 & 0 & 1 \end{bmatrix}\begin{bmatrix} 2 & 3 & 7 \\ 3 & 1 & 2 \\ 1 & 5 & 8 \end{bmatrix}$

28. $\begin{bmatrix} 2 & 3 \\ 1 & 5 \\ 4 & 2 \end{bmatrix}\begin{bmatrix} 1 & 0 \\ 0 & 1 \end{bmatrix}$

29. $\begin{bmatrix} 1 & 0 \\ 0 & 1 \end{bmatrix}\begin{bmatrix} 2 & 3 \\ 1 & 5 \\ 4 & 2 \end{bmatrix}$

In Exercises 30–32, determine whether the following pairs of matrices are inverses to each other.

30. $\begin{bmatrix} 3 & 1 \\ 2 & 1 \end{bmatrix}, \begin{bmatrix} 1 & -1 \\ -2 & 3 \end{bmatrix}$

31. $\begin{bmatrix} 3 & 1 \\ 2 & 1 \end{bmatrix}, \begin{bmatrix} 1 & -1 \\ -2 & 3 \end{bmatrix}$

32. $\begin{bmatrix} 1 & 2 & 3 \\ 2 & 5 & 6 \end{bmatrix}, \begin{bmatrix} -1 & 1 \\ -2 & 1 \\ 2 & -1 \end{bmatrix}$

Solve the following systems of equations using your calculator.

33. $\begin{aligned} x + y &= 2 \\ 3x + 2y &= 5 \end{aligned}$

34. $\begin{aligned} 3x + 7y &= 4 \\ x + 4y &= 5 \end{aligned}$

35. $\begin{aligned} 7x + y &= 1 \\ 3x + 4y &= -1 \end{aligned}$

36. $\begin{aligned} 11x + y &= 3 \\ 3x + y &= 7 \end{aligned}$

37. $\begin{aligned} 3x + 2y &= 8 \\ x + 4y &= 5 \end{aligned}$

38. $\begin{aligned} 15x + 2y &= 10 \\ 21x + 4y &= -13 \end{aligned}$

39.
$$\begin{aligned} 8x + 3y &= 17 \\ -11x + 2y &= 3 \end{aligned}$$

40.
$$\begin{aligned} x - 3y - 2z &= 11 \\ 2x + 8y + 9z &= 23 \\ x + 5y + 6z &= 13 \end{aligned}$$

41.
$$\begin{aligned} x - y + 3z &= 2 \\ 3x - y + 7z &= 3 \\ x + y + 6z &= -5 \end{aligned}$$

42.
$$\begin{aligned} x + 7y + 4z &= -3 \\ -3x - y + 16z &= 7 \\ 2x + 4y - z &= 1 \end{aligned}$$

43.
$$\begin{aligned} x + 3y + 7z &= 4 \\ 3x + 4y + 7z &= 5 \\ 2x + 6y + 9z &= -8 \end{aligned}$$

44.
$$\begin{aligned} 5x + 2y - z &= 4 \\ 4x - 3y + 2z &= 5 \\ 3x + 7y - 5z &= 8 \end{aligned}$$

45.
$$\begin{aligned} 100x + 10y + z &= 13.9 \\ 400x + 20y + z &= 13.5 \\ 900x + 30y + z &= 13.4 \end{aligned}$$

46.
$$\begin{aligned} x + 2z &= -1 \\ 2x + y &= 14 \\ 3y + 5z &= -3 \end{aligned}$$

CHAPTER 8

Modeling Carbon Dioxide Emission from Autos in the United States

8.1 PREDICTING TOTAL PASSENGER CARS IN THE FUTURE

Chart 8.1 below shows the number of U.S. passenger cars (in millions) in the indicated year.

Chart 8.1 Number of Passenger Cars by Year

Years	Cars $\times 10^6$
1940	27.5
1950	40.3
1960	61.7
1970	89.3
1980	121.6
1986	135.4

If we use t for the number of years after 1940, and $A(t)$ for the number of cars ($\times 10^6$) in year $1940 + t$, and plot this data on a (t, A)-coordinate system, it looks like these points fall very nearly in a straight line, so we will model this data with a linear function. See Figure 8.1.

Now, here are some helpful hints on making these reports. If you follow these suggestions, together with whatever else your instructor suggests, you'll probably receive a good grade, and maybe even applause and whistles from your classmates. If you don't do this the first time, you'll probably just have to go back and do it anyway, and maybe get a bad grade, and maybe some boos and catcalls, too, so why not do it right the first time?

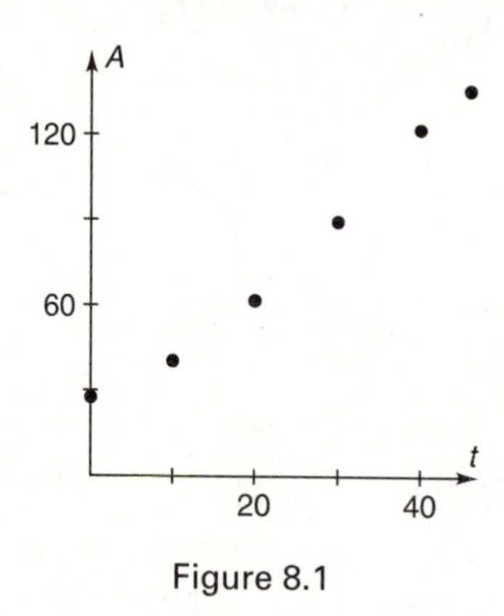

Figure 8.1

You need the best fitting line. Altogether there are fifteen possible lines; your instructor will assign certain years to your group, and your group is to find the best line using points corresponding to these years. Then, each group will report its result.

The first, and most important, thing to remember is this: when you explain what you have done, almost everyone should be able to understand, even if they've not had college algebra. They may not be able to understand the derivation of the equation of the line (or whatever curve), but they should know why you are deriving it, what you intend to do with it, and how it works! More specifically, say what it is that you are modeling; graph the given data; say which years or points you are using to write the equation; show the graph of the equation with the points in the correct position relative to the line (or curve); say what the error is and how your group got it; and finally, show how your equation can be used to predict the future. (You could also impress your class by reporting any additional interesting information your group's equation provides.)

Use your best equation to predict the following information:

1. the number of automobiles in the U.S. in the year 2015;
2. the year in which there will be 200 million automobiles in the U.S.;
3. how many more automobiles there are each year in the U.S.

After each group has reported its best equation, then you will know what the best overall equation is, and you should save it (recycle all the others).

8.2 PREDICTING FUEL EFFICIENCY OF PASSENGER CARS

The next factor that we need to complete our model of CO_2 emission from automobiles is the amount of gasoline they use. The data below (*Statistical Abstracts of the United States*) provide the average gasoline efficiency per car. Chart 8.2 shows the average miles per gallon (MPG) of all U.S. automobiles in the indicated years.

Chart 8.2 Average MPG By Year

Year	MPG
1940	14.8
1950	13.9
1960	13.4
1970	13.5
1980	15.5
1986	18.3

This model is different, so we'll work through the four steps to modeling together. Steps 1 and 2 are easy; Figure 8.2 shows a (t, MPG) coordinate system with the data points plotted (again $t = 0$ in 1940).

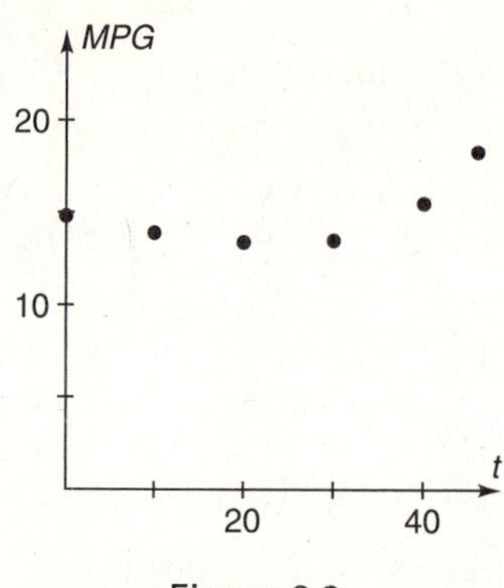

Figure 8.2

Look at the graph and think about what type of curve will best fit this data. After a moment's reflection, you should all agree that these points fall in a very parabolic pattern. Therefore, we will use a quadratic function to model this data, and our job will be to write the equation of the best parabola which fits these points. You may not know how to do this, so think about it for a minute. In particular, think back to your linear equations: you need two points to write the equation of a line, and any two points uniquely determine that line. (There's only one way to draw a line through two points.) There are many ways to draw parabolas through only two points, but there's only one way to draw a parabola through three points. Hence, we must use three points to write the equation of a parabola. Any three points will do. Eventually you'll need to consider all possible triples of data points to get all possible parabolas for this model. We first (somewhat arbitrarily) choose points corresponding to the years 1940, 1970, and 1986, or (0, 14.8), (30, 13.5), and (46, 18.3). To write the quadratic equation whose graph passes through these three points, you need the standard form

$$y = ax^2 + bx + c$$

or, using variables appropriate to our model, the function looks like

$$MPG(t) = at^2 + bt + c,$$

where $MPG(t)$ = average miles per gallon in year 1940 + t. You must always state what your variables mean so that anyone will understand. We'll work out two equations for you, and we put subscripts to distinguish them.

If the graph is to go through the chosen points, then:

when $t = 0$, $MPG_1(0) = 14.8$;

when $t = 30$, $MPG_1(30) = 13.5$;

and when $t = 46$, $MPG_1(46) = 18.3$.

When substituted into the standard form, you get

$$14.8 = a(0)^2 + b(0) + c,$$

$$13.5 = a(30)^2 + b(30) + c,$$

$$18.3 = a(46)^2 + b(46) + c,$$

so you have to solve the system of three linear equations in the unknowns a, b, and c:

$$14.8 = c$$

$$13.5 = 900a + 30b + c$$

$$18.3 = 2116a + 46b + c.$$

The coefficient matrix is

$$A = \begin{bmatrix} 0 & 0 & 1 \\ 900 & 30 & 1 \\ 2116 & 46 & 1 \end{bmatrix}$$

and the matrix of constants is

$$B = \begin{bmatrix} 14.8 \\ 13.5 \\ 18.3 \end{bmatrix}.$$

The product matrix $A^{-1}B$ is $\begin{bmatrix} 0.0075 \\ -0.2672 \\ 14.8000 \end{bmatrix}$, and hence the solution to this system (see Chapter 7) is

$$a = 0.0075$$

$$b = -0.2672$$

$$c = 14.8000.$$

The corresponding quadratic function is

$$MPG_1(t) = 0.0075t^2 - 0.2672t + 14.8$$

and its graph is shown in Figure 8.3.

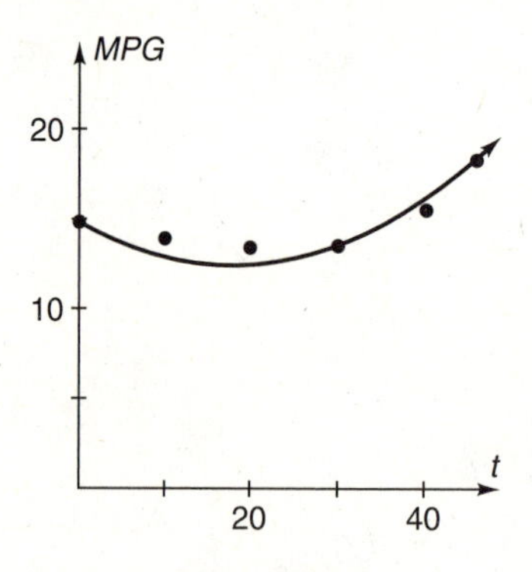

Figure 8.3

Notice that the parabola actually passes through the chosen points (0, 14.8), (30, 13.5), and (46, 18.3). Also notice that the original data point (40, 15.5) is below the parabola; that's because

$$MPG_1(40) = 16.11,$$

which is larger than 15.5.

One more example to be sure you've got the hang of it. This time use the points (10, 13.9), (20, 13.4), and (30, 13.5). The system of equations is

$$13.9 = 100a + 10b + c$$

$$13.4 = 400a + 20b + c$$

$$13.5 = 900a + 30b + c.$$

The coefficient matrix of the system is:

$$A = \begin{bmatrix} 100 & 10 & 1 \\ 400 & 20 & 1 \\ 900 & 30 & 1 \end{bmatrix},$$

and the matrix of constants is

$$B = \begin{bmatrix} 13.9 \\ 13.4 \\ 13.5 \end{bmatrix}.$$

The product matrix A^{-1}B is $\begin{bmatrix} 0.003 \\ -0.140 \\ 15.000 \end{bmatrix}$.

The corresponding quadratic function is

$$MPG_2(t) = 0.003t^2 - 0.140t + 15.$$

Its graph is shown in Figure 8.4.

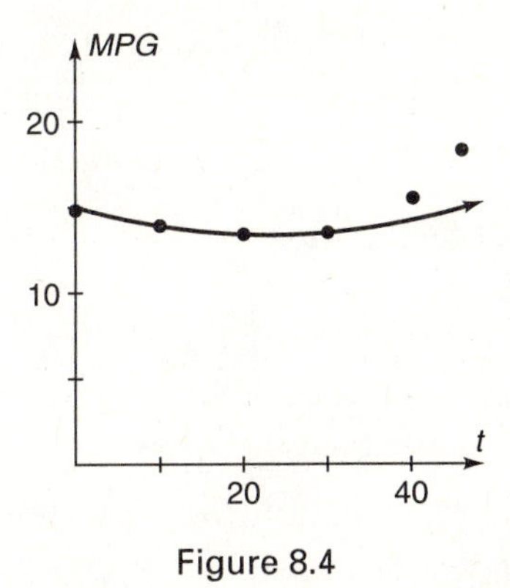

Figure 8.4

Now, we have derived two of the possible twenty quadratic functions which could serve as a model for this data. Of course, we must find the best one, and this involves finding the error for each. The errors for parabolic models are computed the same way as the errors for linear models, and the function with smallest error is the best. Chart 8.2B on page 96 shows the actual data values for the relevant years along with the functional values (predicted values) for each of the three derived quadratic functions.

So the best of these two functions is $MPG_1(t)$. Even though this may not (it isn't) be the best function which models this data, we will use it to illustrate the type of information such an equation can provide.

Chart 8.2B

t	Actual	$MPG_1(t)$	$MPG_2(t)$	Difference for: MPG_1	Difference for: MPG_2
0	14.8	14.8	15.0	0.0	0.2
10	13.9	12.9	13.9	1.0	0.0
20	13.4	12.5	13.4	0.9	0.0
30	13.5	13.5	13.5	0.0	0.0
40	15.5	16.1	14.2	0.6	1.3
46	18.3	18.4	14.9	0.1	3.4
			Error	2.6	4.9

EXAMPLE 8.1

In order to determine what the average fuel efficiency will be in 1995, you first need to determine t. The variable t represents the number of years after 1940, so $t = 55$. Next, use your calculator to evaluate the function at 55, i.e.,

$$MPG_1(55) = 22.8 \text{ mpg.}$$

EXAMPLE 8.2

Many environmentalists are lobbying for a fuel efficiency of 45 mpg by the year 2000. Suppose that automobile manufacturers do not yield to this pressure, and that the trend shown by our function continues. In what year would fuel efficiency actually reach 45 mpg? To decide this, you need to solve the quadratic equation

$$45 = MPG(t),$$

$$45 = 0.0075t^2 - 0.2676t + 14.8.$$

Put this into standard form:

$$0 = 0.0075t^2 - 0.2676t - 30.2,$$

and use the quadratic formula to get

$$t = \frac{-(-0.2676) \pm \sqrt{(-0.2676)^2 - 4(0.0075)(-30.2)}}{2(0.0075)},$$

or $t = -48.1$ and $t = 83.8$. Should $t = -48.1$, then the year would have been

$$1940 - 48 = 1892.$$

Of course, there were very, very few cars then! So the answer is $t = 83.8$, or 84 rounded. The year will be $1940 + 84 = 2024$, and that's twenty-four years after the environmentalists' target year of 2000.

If you use your calculator graph to solve this equation, you would use the trace feature to estimate the first coordinate when the second coordinate is approximately 45. See if you can determine when fuel efficiency will reach 30 mph by using the trace feature. The answer is $t = 66$, or in the year $1940 + 66 = 2006$.

EXAMPLE 8.3

When do you think fuel efficiency was poorest? This involves finding the vertex of the parabola. Remember, when a parabola opens up, the second coordinate of its vertex is its minimum value. The first coordinate is given by

$$t = -\frac{b}{2a} = \frac{-(-0.2676)}{2(0.0075)} = 17.84.$$

This corresponds to the year 1958. The gas mileage was only

$$MPG_2(18) = 12.4 \text{ mpg}.$$

Not too great. Some of you may remember, or have seen pictures of, those huge Cadillacs with the big fins—major contributors to atmospheric CO_2. But, weren't they spectacular!

Figure 8.5 shows the graph of $MPG_2(t)$ with relevant points indicated.

Figure 8.5

It is your task to come up with the best parabola for this data. You should do this in your group after your instructor has assigned years to you. There will be a total of eighteen remaining equations. Each group representative should report her or his best parabola to the class. Also, show its graph, do some predictions for future years, and answer this question: "When was gasoline efficiency at its worst?

Once the reports are done, you will know the best function for gas mileage; call it $MPG(t)$; save it for future use.

Discuss what social, political, or physical changes might effect the accuracy of this model.

The following is an interesting study on gasoline and carbon dioxide.

1. What is the gas mileage on your car?
2. How many gallons of gasoline does your car burn in one mile?
3. If average fuel efficiency improves each year, cars will burn less gasoline per mile. Use the equation for $MPG(t)$ derived in this section to write the function $GPM(t)$ which defines gallons of gasoline burned each mile in terms of year $1940 + t$.

4. Enter this function $GPM(t)$ into your calculator and graph. Use "trace" to determine its maximum, and explain what this means.
5. Predict the average number of gallons of gasoline which will be burned per mile per car in the year 1995.
6. Each gallon of gasoline burned emits 20 pounds of CO_2. Use your answer to Exercise 2 above to determine the number of pounds of CO_2 your car emits each mile you drive. Estimate the total number of miles you drive each year, and find out how many pounds of CO_2 your car emits in a year.
7. Use your answer to Exercise 5 above to predict the average number of pounds of CO_2 which will be emitted per mile per car in the year 1995.

Use the following information to complete the methanol study below.

i) It takes 75% more methanol than gasoline to go one mile.

ii) When burned, one gallon of methanol emits 9.5 pounds of CO_2.

8. Use your answer to Exercise 3 in the above gasoline study to determine the function which defines gallons of methanol which would be burned each mile in terms of year $1940 + t$.
9. Use your answer to the preceding problem to predict the average number of gallons of methanol which would be burned per mile per car in the year 1995.
10. Use your answer to the preceding problem to predict the average number of pounds of CO_2 per mile per car which would be emitted in the year 1995.

Answers to Exercises 5, 7, 9, and 10 will be needed in Chapter 20.

8.3 AVERAGE YEARLY MILEAGE OF CARS

There's a great line in a great movie called "Repo Man": "The more you drive, the less intelligent you are." If you haven't seen the movie, check it out.

Should be at your favorite video store. Chart 8.3 indicates the average number of miles each automobile in the United States is driven in the designated year.

Chart 8.3

Year	Avg. Miles Driven per Car per Year ($\times 10^3$)
1940	9.1
1950	9.1
1960	9.5
1970	10.0
1980	8.8
1986	9.3
1987	9.6

Take $t = 0$ to be 1940 and plot the corresponding points on a (t, M) coordinate system, where

$$M(t) = \text{average miles driven (in thousands) per car in year } 1940 + t.$$

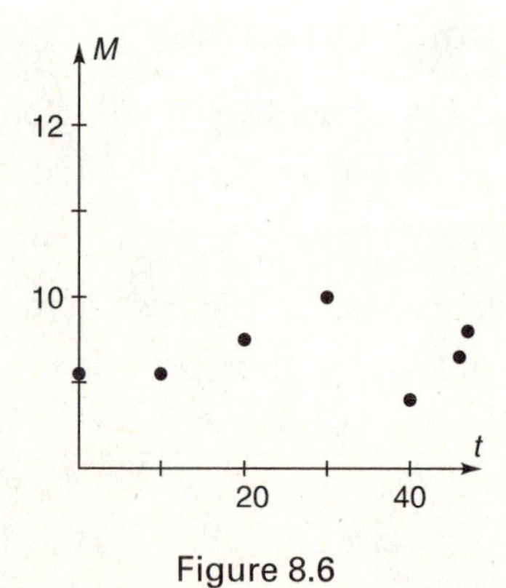

Figure 8.6

Figure 8.6 shows these points on the graph. The pattern made by these points is probably unfamiliar to you—it certainly doesn't look like any ONE of the functions listed as possibilities for use as a model in *Earth Algebra*. The key word in the last sentence is ONE. Now, before reading on, look at the figure and think about how you might model this data.

What we see (we, meaning the authors) is not just one curve, but two. Maybe some of you saw a couple of lines, one from 1940 to 1970 and the other from 1980 on into the future. The problem with using two lines is that you would end up with a big gap in your overall graph—neither good for looking back into the past, nor describing continuous trends. The data is best modeled with two curves, the first going from 1940 to 1970, and the second from 1970 on into the future. These curves should have a common point in the year 1970 to avoid the discontinuity mentioned above. The data for 1940–1970 possibly could be approximated with a line, but a parabola is a better fit (why?). So we choose to model this data with two parabolas.

This kind of function is called a *piecewise function*: "piecewise" because its definition changes over different time intervals, although you still have only one function. For this model, we have this requirement: the parabola for 1940–1970 must have the point corresponding to 1970 at its right hand endpoint, and the parabola which starts in 1970 and goes on into the future must have the same point for its left hand endpoint. In other words, the first parabola ends at the same point where the second one begins—this avoids the big gap in the graph.

More specifically, you have two sets of data points. The first set is in Chart 8.4A; the second set is in Chart 8.4B. This is just the data in Chart 8.3 broken into two parts, with 1970 common to each.

Chart 8.4A

Year	Miles per Car
1940	9.1
1950	9.1
1960	9.5
1970	10.0

Chart 8.4B

Year	Miles per Car
1970	10.0
1980	8.8
1986	9.3
1987	9.6

The task is to find the best parabola which models the data in Chart 8.4A, and then to find the best one which models the data in Chart 8.4B. The equation for any parabola for 1940–1970 must pass through the point (30, 10.0), and the same goes for any parabola for 1970–future.

After all this is done, you have two equations, but only one function which models the data in Chart 8.3. This is a piecewise function. Here is an example. One quadratic equation for 1940–1970 is

$$M_1(t) = .0005t^2 + .025t + 8.8 \text{ (subscript 1 for first interval)},$$

which is derived using points (10, 9.1), (20, 9.5), and (30, 10.0). It is good for approximating data from 1940–70 only. A quadratic equation for 1970–1987 is

$$M_2(t) = 0.0202t^2 - 1.58t + 39.2,$$

which is derived using points (30, 10.0), (46, 9.3) and (47, 9.6). But this one can only be used for the years 1970 on. Finally, the ONE function is the piecewise function $M(t)$ defined by

$$M(t) = \begin{cases} 0.0005t^2 + 0.025t + 8.8 & \text{if } 0 \le t \le 30, \\ 0.0202t^2 - 1.58t + 39.2 & \text{if } t \ge 30, \end{cases}$$

and its graph is shown in Figure 8.7.

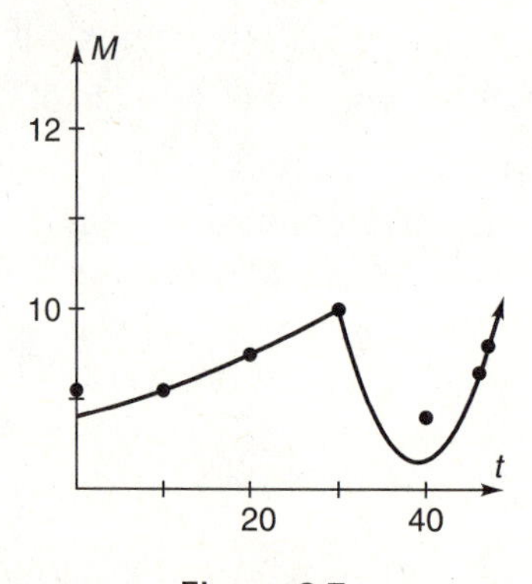

Figure 8.7

Write all possible quadratic equations which model the data in Chart 8.4A—you must always use as one point (30, 10.0). When you finish writing all such equations (there are only three), find the error using only points corresponding to 1940–1970. The equation with the smallest error is the best one for the time interval 1940–1970.

Next, repeat the same exercise for years from 1970–1987. Again, you must always use (30, 10.0) as one point. Find the one with the smallest error; this is the one to use for years 1970 on.

Use your model to answer the following questions:

1. Estimate the number of miles each car drove in 1953.
2. Estimate the number of miles each car will drive in 1998.
3. When will the average number of miles driven be 12,000?
4. When was the average number of miles driven 9,700?

Discuss possible limitations of this model.

CHAPTER 9

Modeling Carbon Dioxide Emission from Power Consumption in the United States

In this chapter energy consumption data in the United States are provided (*Statistical Abstracts of the United States*). Fuels are coal, petroleum, and natural gas, and quantities are measured in quadrillion BTU (quadrillion means 10^{15}), abbreviated "quads." Each set of data is to be modeled by the groups, and finally total energy consumption will be determined.

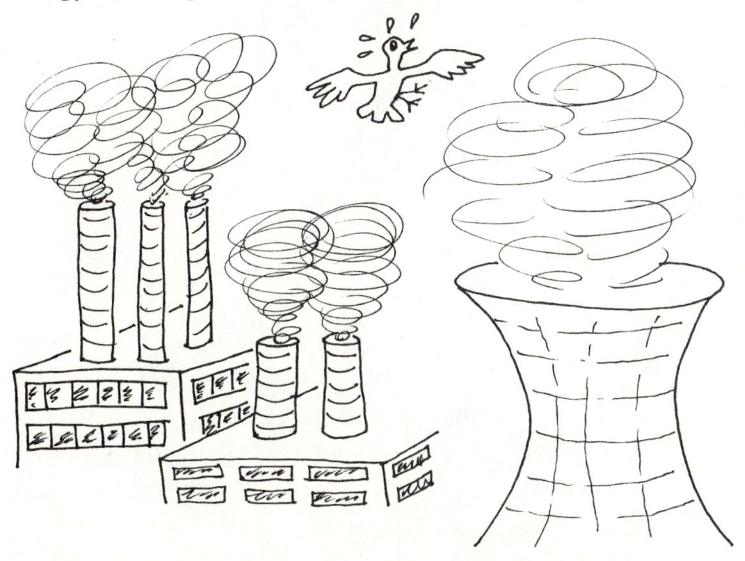

9.1 PREDICTING COAL CONSUMPTION

Here are the figures for coal consumption in the U.S. (Chart 9.1):

Chart 9.1

Year	Coal Consumption (quads)
1970	12.28
1975	12.62
1980	15.43
1985	17.46
1988	18.78

Figure 9.1 shows these data plotted on a $(t, CC(t))$ coordinate system where $CC(t)$ is coal consumption (in quads) in year $1970 + t$. Since all of the information starts with 1970, we let $t = 0$ correspond to 1970.

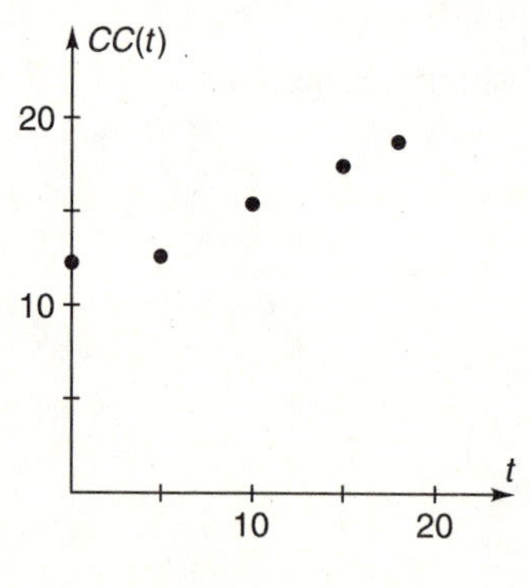

Figure 9.1

The earlier points fit nicely into a parabola, but the later three points appear to be nearly collinear. Therefore, we suggest a piecewise function as a model: a parabola for years 1970–1980, and a line for years 1980–future. Remember, all possible lines for the later years must begin at the common point, in 1980, in order to preserve continuity.

It is a task for the class to determine the best model for coal consumption. Your final model will be only ONE function, although it is defined by two different equations.

Use your model to obtain the following information.

1. Predict the coal consumption for the year 1997.
2. What was coal consumption in 1976?
3. When will coal consumption be 25 quads?

9.2 PREDICTING PETROLEUM CONSUMPTION

Data for consumption of petroleum are in Chart 9.2.

Chart 9.2

Year	Petroleum Consumption (quads)
1970	29.48
1975	32.71
1980	34.20
1985	30.93
1988	33.96

The corresponding points are plotted on a $(t, PC(t))$ coordinate system (Figure 9.2) with $t = 0$ for 1970, and $PC(t)$ = petroleum consumption (in quads) in year $1970 + t$.

Once again, do you see two different patterns? You can use a piecewise curve again, but this time use a linear function for the first (years 1970–1980), and a quadratic for the second (1980–future).

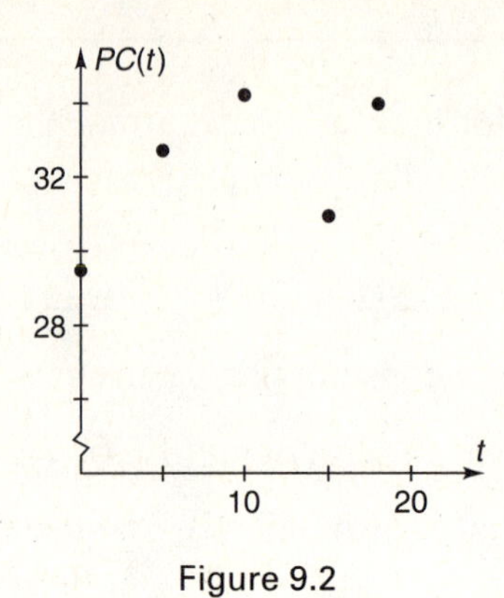

Figure 9.2

The class should now derive the best model for petroleum consumption. Use your model to derive the following information.

1. Predict petroleum consumption in year 2001.
2. What was petroleum consumption in 1973?
3. When was petroleum consumption 33 quads?
4. When is consumption equal to 40 quads?

9.3 PREDICTING NATURAL GAS CONSUMPTION

Data for consumption of natural gas are in Chart 9.3.

Chart 9.3

Year	Natural Gas Consumption (quads)
1970	21.78
1975	19.95
1980	20.44
1985	17.83
1988	18.62

The points corresponding to the data in Chart 9.3 are plotted in Figure 9.3 on a $(t, NGC(t))$ coordinate system, where $NGC(t)$ = natural gas consumption (in quads) in year $1970 + t$.

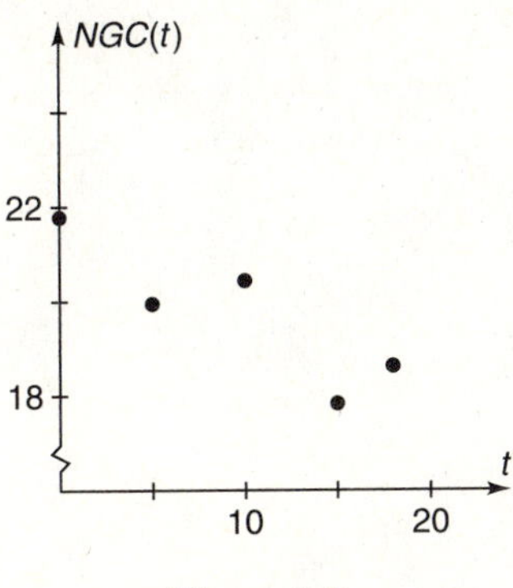

Figure 9.3

This suggests (we hope) another piecewise curve, in this case two parabolas with 1980 serving as the common year.

The class should derive the best model for natural gas consumption. Use your model to determine the following.

1. Predict natural gas consumption in 2010.
2. What was consumption in 1976?
3. In what year(s) will natural gas consumption once again reach the 1970 level?
4. When was consumption least? What was the amount?

9.4 PREDICTING TOTAL ENERGY CONSUMPTION

The patterns for total energy consumption are more complicated since they involve more factors; included in the total consumption figures are other sources of power, such as solar, nuclear, wood burning, and geothermal. It seems that for the three energy sources we studied, patterns dramatically changed in 1980, so for our total consumption model, we concentrate only on data obtained for years from 1980 through 1988.

Chart 9.4

Year	Total Energy Consumption (quads)
1980	76.0
1982	70.8
1985	74.0
1987	76.8
1988	79.9

group work

Plot all the data points on a (t, TEC) coordinate system, where $TEC(t)$ is total energy consumption (in quads) in year $1970 + t$. We think you will agree that a parabola appears to be a good model. The class should now determine the best parabola to use to model the data.

Use your model to determine the following.

1. In what year was energy consumption the least?
2. Predict energy consumption for the year you plan to graduate from college.
3. In what year will energy consumption reach 120 quadrillion BTU?

Discuss social, political or physical changes that might effect the accuracy of your models.

Save all models derived in this chapter. You'll need them later.

CHAPTER 10

Exponential and Logarithmic Functions

10.1 EXPONENTIAL FUNCTIONS

The general form we use for an *exponential function* is

$$f(x) = a(b^x),$$

where a and b are constant, and x is the variable. The variable x is called the *exponent*, and the constant b is called the *base*. Restrictions on b are required: b must be positive and different from 1. (Note that if $b = 1$, then this function would be constant, and the restriction of b to positive values allows the domain of an exponential function to be all real numbers x.) Regardless of the value of x, $b^x > 0$, and hence the range of an exponential function will be all $y > 0$ if $a > 0$, and all $y < 0$ if $a < 0$.

Exponential functions are easily graphed on your calculator; here are some examples.

Note: Chapter 10 is a prerequisite for Chapters 11 and 14.

EXAMPLE 10.1

Graph $f(x) = 2^x$ on your calculator. (Here, $a = 1$ and $b = 2$.)

Enter this function into your calculator, and graph. You should see the graph as shown in Figure 10.1. The "range" for this graph is set as follows:

x min = –2; x max = 2; y min = 0; y max = 10.

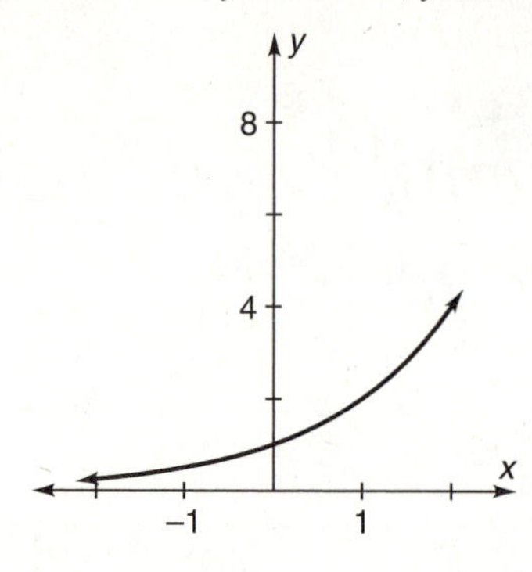

Figure 10.1

You should note that there are no x-intercepts; the y-intercept is (0, 1) which means $2^0 = 1$, and the entire graph is above the x-axis which means that $2^x > 0$ for all x.

EXAMPLE 10.2

Leave the function $f(x) = 2^x$ in your calculator, and enter this second function

$g(x) = 10^x$.

(Here $a = 1$ and $b = 10$.) Now press **graph**, and you should see the graph of both exponential functions on your screen (Figure 10.2).

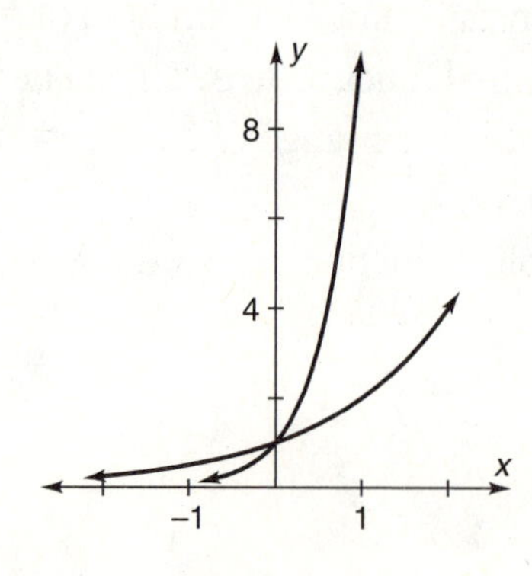

Figure 10.2

These graphs look basically the same, but $y = 10^x$ is a lot steeper than $y = 2^x$. In general, the larger the base, the steeper the graph, which means that the function increases really fast when the base is big.

EXAMPLE 10.3

Graph $h(x) = (0.5)^x$ on your calculator. Now $b = 0.5 < 1$ and the graph should appear as in Figure 10.3. This graph is decreasing, whereas the graph of $y = 2^x$ is increasing. If the base b is smaller than 1, the exponential function $y = b^x$ decreases, and if the base b is larger than 1, $y = b^x$ increases. Also, note from Figure 10.3 that $(0.5)^x > 0$.

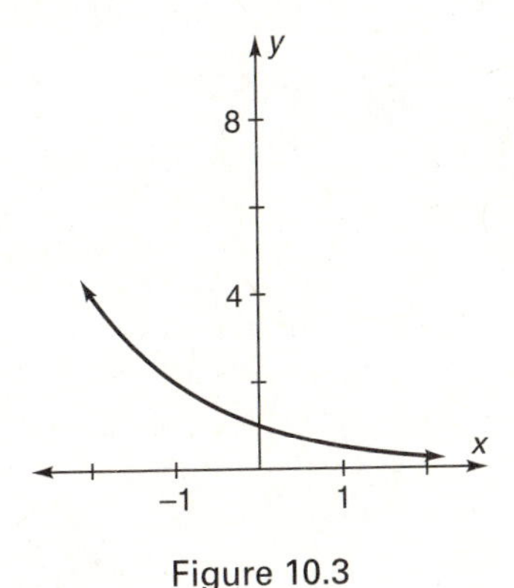

Figure 10.3

We list some properties which are true for all exponential functions of the form $y = b^x$.

1. There are no x-intercepts for any x.
2. The y-intercept is (0, 1): $b^0 = 1$.
3. The entire graph is above the x-axis ; $b^x > 0$ for all x.
4. If $0 < b < 1$, the function decreases.
5. If $b > 1$, the function increases.

Properties 6–9 are the ones which are commonly known as the "laws of exponents."

6. $b^p \times b^q = b^{p+q}$.
7. $\dfrac{b^p}{b^q} = b^{p-q}$.

8. $(b^p)^q = b^{pq}$.
9. $b^{-p} = \dfrac{1}{b^p}$.

EXAMPLE 10.4

Graph $y = 0.53(1.07)^x$ on your calculator, and determine the y-intercept.

Enter this function into your calculator and graph (Figure 10.4). This example is slightly different from the first three in that

$$a = 0.53;$$

this makes the y-intercept (0, 0.53).

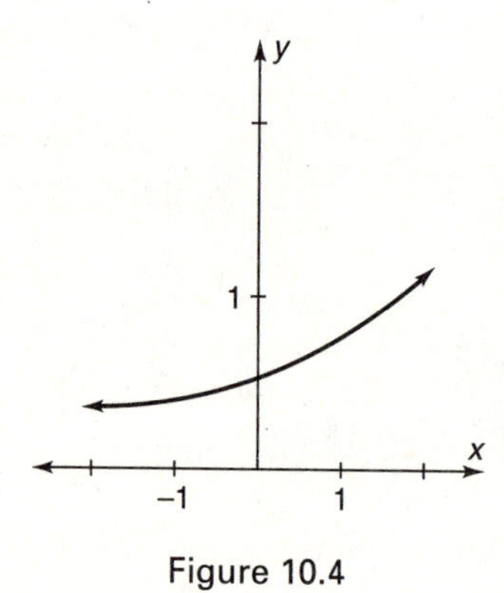

Figure 10.4

EXAMPLE 10.5

A very special number in mathematics is known as "e" with approximate value 2.71828. This number e occurs quite naturally in a variety of studies, in particular, in physics, biology, and even economics. The exponential function

$$y = e^x$$

is known as the *natural exponential function*, its base is

$$b = e,$$

and its graph is shown in Figure 10.5.

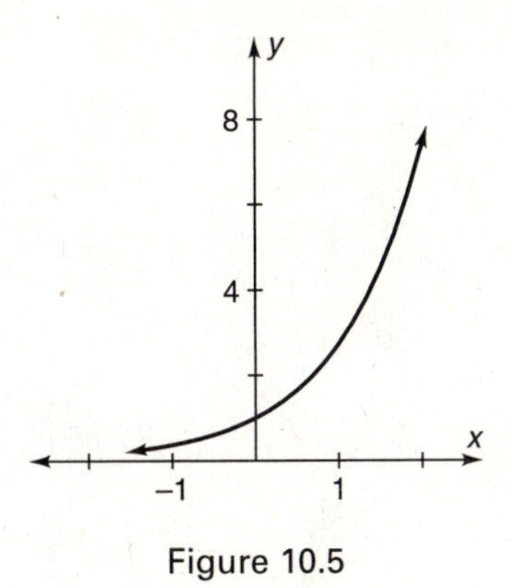

Figure 10.5

Turn to Section 10.4 and solve Problems 1–10.

10.2 LOGARITHMIC FUNCTIONS

The notation for the logarithmic function is $\log_b x$; this is read: "*the logarithm with base b of the number x*". The definition of this function is:

$$y = \log_b x \text{ if } b^y = x.$$

The exponent y in the expression $b^y = x$ can take on any real value y (positive, negative, or zero), and so the range of this logarithmic function consists of all real numbers. But $x = b^y$ is always positive, so the domain of the logarithmic function consists of only positive numbers.

Recall that the base b can change, but must always be a positive number different from 1. Obviously, this logarithmic function is closely related to an exponential function (see Section 16.3). You get a different logarithmic function whenever you change the base b. Some examples should help your understanding of this function.

EXAMPLE 10.6

Determine $\log_2 8$.

The base here is 2 so

$$\log_2 8 = y \text{ if } 2^y = 8.$$

What power do you need to take 2 to in order to get 8? The answer is, of course, 3, since $2^3 = 8$. So

$$\log_2 8 = 3.$$

EXAMPLE 10.7

Determine $\log_{10} 100$.

Here $\log_{10} 100 = y$ if $10^y = 100$. The answer is 2 because $10^2 = 100$, so

$$\log_{10} 100 = 2.$$

In this example, if we use the functional notation $f(x) = \log_{10} x$, then

$$f(100) = 2.$$

EXAMPLE 10.8

Determine $\log_{10} 0.1$.

You have to do a little more work here. You must find the exponent in the equation

$$10^y = 0.1.$$

The number 0.1 is the same as $\frac{1}{10}$, which is the same as 10^{-1}, so you have the equation

$$10^y = 10^{-1},$$

so $y = -1$; i.e., $\log_{10} 0.1 = -1$.

We now list the properties of logarithms. These will be needed later when we solve both exponential and logarithmic equations.

1. $\log_b (MN) = \log_b M + \log_b N$.
2. $\log_b \frac{M}{N} = \log_b M - \log_b N$.
3. $\log_b (M^p) = p \log_b M$.
4. $\log_b b = 1$.
5. $\log_b 1 = 0$.
6. $\log_b b^x = x$.
7. $b^{\log_b x} = x$.

These properties can be derived from corresponding properties of exponents.

EXAMPLE 10.9

Graph $y = \log_{10} x$.

The function $\log_{10} x$ is called the *common logarithm*, and it is usually abbreviated as $\log x$. Your calculator should have a key labeled **LOG**, and so to determine the graph, just enter the function $y = \log x$ and press **graph**. Figure 10.6 shows the graph with

$$x \min = 0,\ \ x \max = 10,\ \ y \min = -2,\ \ y \max = 2.$$

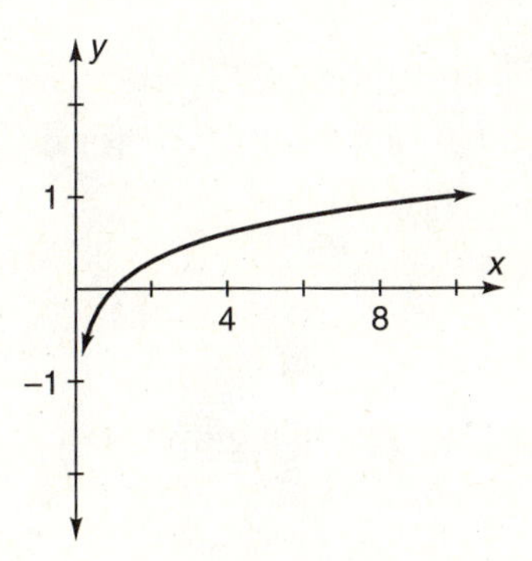

Figure 10.6

The function $\log_e x$ is called the *natural logarithmic function*, and is written $y = \ln x$. We are going to concentrate on natural logarithms since they are the most useful in *Earth Algebra*. The graph of $y = \ln x$, which looks very much like $y = \log x$, is shown in Figure 10.7.

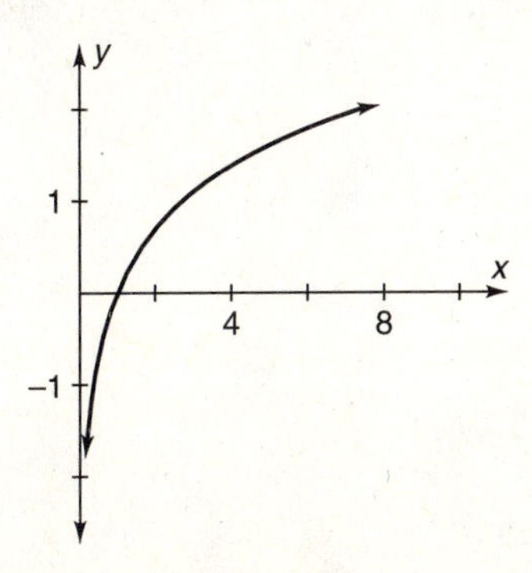

Figure 10.7

The domain of $y = \ln x$ consists of all $x > 0$, and the range is all real numbers y. The graph doesn't go to the left of the y-axis; it increases from $-\infty$ to $+\infty$, but it increases very, very slowly. Its x-intercept is (1, 0), because $e^0 = 1$ and so $\ln 1 = 0$; it has no y-intercept. The function $y = \ln x$ is easy to evaluate on your calculator. For example, to find ln 3.1 press the **ln** key, then 3.1, then enter:

$$\ln 3.1 = 1.1314.$$

Note that $\ln e = 1$, since $e^1 = e$.

The standard form for the logarithmic function which will be used in *Earth Algebra* is

$$y = a + b \ln x,$$

where a and b are constants. We now look at some graphs (on the calculator) of logarithmic functions of this type for different values of a and b.

EXAMPLE 10.10

Graph $y = 3 + 2 \ln x$.

Figure 10.8 shows this graph.

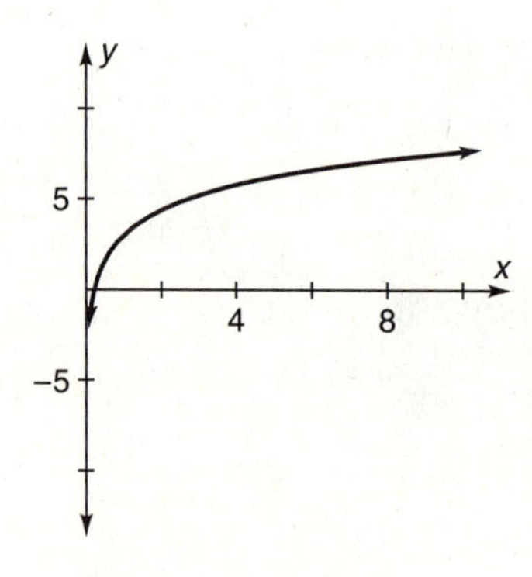

Figure 10.8

EXAMPLE 10.11

Graph $y = 2.5 \ln x - 1.7$.

Figure 10.9 shows this graph.

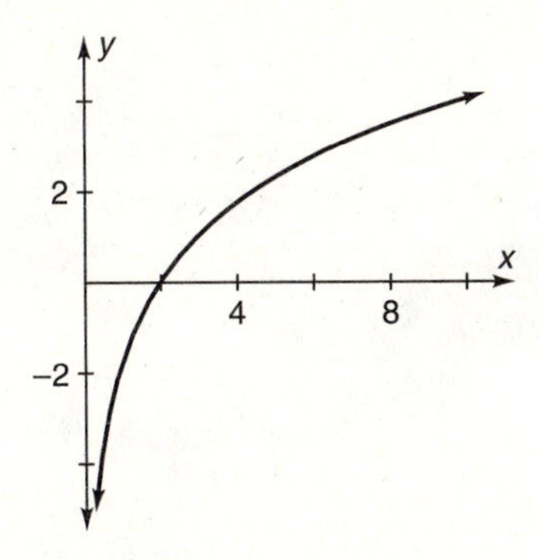

Figure 10.9

Turn to Section 10.4 and solve Problems 11–22.

10.3 LOGARITHMIC AND EXPONENTIAL EQUATIONS

Consider Example 10.11 in the preceding section. To estimate the x-intercept of this graph, use the trace feature on the calculator to get $x = 1.97$. To find this intercept algebraically, it is necessary to solve the equation

$$0 = 2.5 \ln x - 1.7.$$

First solve for $\ln x$:

$$1.7 = 2.5 \ln x$$

$$\ln x = 0.68.$$

Therefore $e^{\ln x} = e^{0.68}$, and recalling from Property 7 of logarithms that $e^{\ln x} = x$, we have

$$x = e^{0.68} = 1.974.$$

EXAMPLE 10.12

Let $f(x) = 1.6 \ln x + 4$, and algebraically determine the x-intercept of its graph.

Solve the equation

$$0 = 1.6 \ln x + 4$$

$$-4 = 1.6 \ln x$$

$$\ln x = -2.5.$$

Therefore

$$e^{\ln x} = e^{-2.5}$$

$$x = e^{-2.5} = 0.082.$$

EXAMPLE 10.13

Let $f(x) = 6 + 8.1 \ln x$. Determine x if $f(x) = 4$.

Solution A. Estimate the x value when $f(x) = 4$ using the **trace** on your calculator.

To do this, graph $y = 6 + 8.1 \ln x$ and the horizontal line $y = 4$ on your calculator (Figure 10.10).

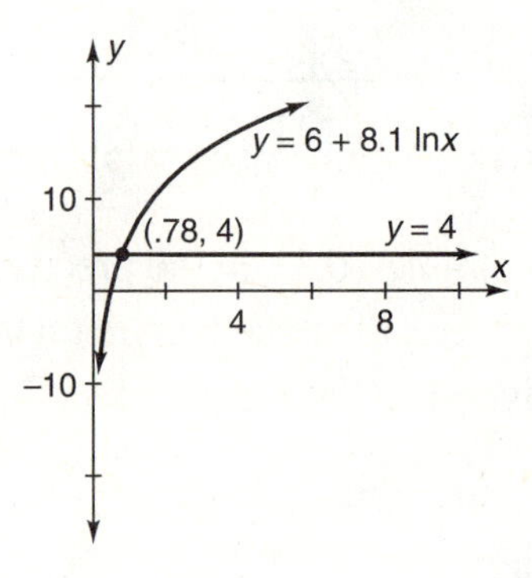

Figure 10.10

The x value we want to estimate is the x-coordinate of the point of intersection of these graphs. Use **trace** to find

$$x = .78.$$

Solution B. Solve for x algebraically when $f(x) = 4$.

We need to solve the equation

$$4 = 6 + 8.1 \ln x$$

$$-2 = 8.1 \ln x$$

$$\ln x = -0.2469,$$

so $x = e^{-0.2469} = 0.7812$.

Note that the point of intersection of the graphs is (0.7812, 4).

EXAMPLE 10.14

Solve the exponential equation

$$e^{2x+3} = 7.$$

Let $f(x) = e^{2x+3}$.

Solution A. Estimate the x value when $f(x) = 7$. Graph $y = e^{2x+3}$ and the horizontal line $y = 7$ on your calculator (Figure 10.11). The x value we want to estimate is the x-coordinate of the point of intersection of these graphs. Use **trace** to find

$$x = -0.53.$$

Solution B. Solve for x algebraically when $f(x) = 7$.

$$7 = e^{2x+3}.$$

Take the natural logarithm of each side to get

$$\ln 7 = \ln (e^{2x+3}).$$

Recall from Property 6 that $\ln (e^z) = z$, so we get

$$\ln 7 = 2x + 3.$$

Next evaluate ln 7 on the calculator and solve for x:

$$1.9459 = 2x + 3$$

$$2x = -1.0541$$

$$x = -0.5270.$$

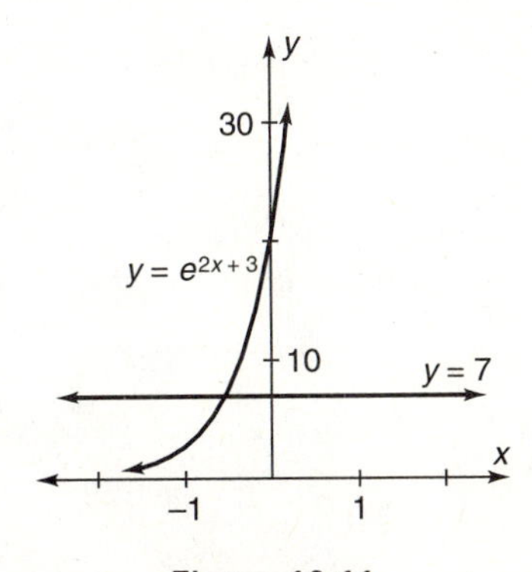

Figure 10.11

EXAMPLE 10.15

Solve the exponential equation

$$2(1.2^x) = 5.$$

First divide by 2 to isolate the "exponent part."

$$(1.2^x) = 2.5.$$

Next take the natural logarithm of each side:

$$\ln (1.2)^x = \ln (2.5);$$

then recall Property 3 of logarithms which says that $\ln b^x = x$. This gives

$$x \ln (1.2) = \ln (2.5);$$

Now we can solve for x:

$$x = \frac{\ln (2.5)}{\ln (1.2)} = 5.0257.$$

Solve Problems 23–28 in the next section.

10.4 THINGS TO DO WITH EXPONENTS AND LOGARITHMS

Use your calculator to evaluate each of the following.

1. $10^{.259}$.
2. $(1.72)^{5.61}$.
3. e^2.
4. $3e^{2.1}$.
5. $7.31(5.6)^{9.74}$.

Graph each of the functions in 6–10 on your calculator and determine the y-intercept.

6. $f(x) = 6^x$.
7. $g(x) = -3^x$.
8. $h(x) = (1.13)^x$.

9. $k(x) = 9.6(4.3)^x$.

10. $l(x) = -2.4e^x$.

For 11–17 determine the logarithms; use your calculator for 14–17.

11. $\log_3 81$.

12. $\log_{10} .0001$.

13. $\log_{10} 10000$.

14. $\log 17$.

15. $\log .023$.

16. $\ln 2$.

17. $\ln .03$.

For 18–22, graph the given equation; estimate the x-intercept for each.

18. $y = 3 + 5 \log x$.

19. $y = 5 - 2 \log x$.

20. $y = 2 \ln x$.

21. $y = 1 - 3.2 \ln x$.

22. $y = 500 + 2.3 \ln x$.

A. graph each function on your calculator and use the graph to solve the given equation. B. Solve each equation algebraically.

23. $f(x) = 1.2 + 3.7 \ln t$; solve $1.2 + 3.7 \ln t = 9$.

24. $y = 500 + 225 \ln x$; solve $500 + 225 \ln x = 100$.

25. $H(s) = .002 - .035 \ln s$; solve $.002 - .035 \ln s = .04$.

26. $g(x) = e^{x+1}$; solve $e^{x+1} = 2$.

27. $r(t) = e^{8t-4}$; solve $e^{8t-4} = 1$.

28. $p(t) = 1.7(3^t)$; solve $1.7(3^t) = 4.5$.

CHAPTER 11

Modeling Carbon Dioxide Emission Resulting from Deforestation

11.1 PREDICTING DEFORESTATION DUE TO LOGGING

Governments of developing countries are ready and willing to export forest products, particularly timber, in order to obtain currency from foreign countries. Large timber companies from North America and Japan are happy to collect profits from imported rain forest woods. Not only does this contribute to global warming, but it also depletes supply for local consumption, in particular, for heating and cooking. Essentially, poor forestry practice is harmful for all parties involved.

The information provided below (Chart 11.1) is the estimated number of hectares (ha) of rain forest lost to logging by year.

Chart 11.1 Rain Forest Cut for Logging

Year	ha × 10^6
1960	.44
1970	1.19
1980	2.20
1988	3.90

Groups should model this data using $t = 0$ in 1940. (Later you will see why we have chosen 1940 instead of 1960.) We think you will agree that this looks parabolic. Your instructor will assign years to each group. When your model is complete, name your best function $L(t)$, and preserve it.

Use your model to determine the following information.

1. How many hectares will be cut for logging in the year 2001?
2. There are 11.6×10^6 ha of ancient forests in the U.S. In what year will the area of rain forest cut for logging alone equal this number?

11.2 PREDICTING DEFORESTATION DUE TO CATTLE GRAZING

The destruction of a major portion of the rain forests in Central and South America has been to provide grazing lands for cattle. Cattle ranching has been encouraged by local governments, and most of the resulting beef went to the United States for use at fast food restaurants and for pet foods. Due to inefficiency, cattle ranching has decreased recently; indeed, most rain forest soil does not provide good pasture, and has limited use. It takes approximately two and one-half acres to feed one cow, and in addition to that, erosion and compactification render the land unusable after only a few years.

The data provided below (Chart 11.2) is the estimated amount of rain forest cleared for cattle ranching in the indicated year.

Chart 11.2 Rain Forest Cut for Cattle Grazing

Year	ha $\times 10^6$
1960	0.30
1970	0.95
1980	1.36
1988	1.40

The graph (Figure 11.1) shows these points plotted on a (t, CG) coordinate system with $t = 0$ in 1940.

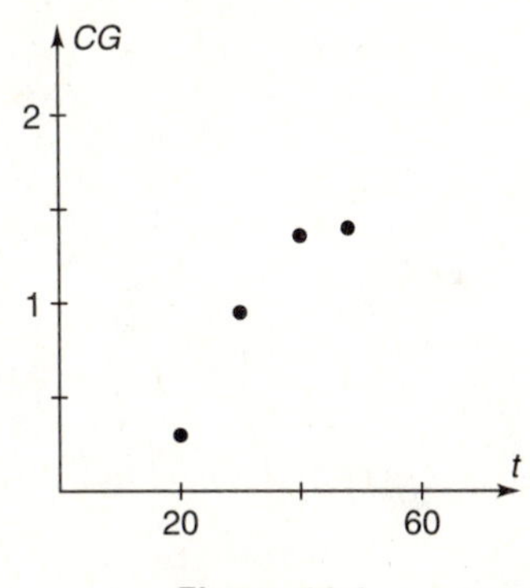

Figure 11.1

We need to discuss the type of equation to use to model this data. Note that the number of hectares cut is increasing from 1950 through 1988, but the rate of increase is slowing from 1950 through 1988 (slopes between consecutive data points are .065, .041, and .018). Hence a linear model would not be good. Your next guess might be a parabolic model. This would be O.K. except for one problem: the vertex would occur in some year in the future, which would indicate that the amount of land cleared for cattle would reach a maximum and then start decreasing after that year. Without any information to support such an event, this could give very inaccurate predictions. Therefore, it would be nice to have an equation which more accurately reflects this decreasing rate of increase, but with no vertex in the future. What curve do you know that behaves like that? The one that we will use is a logarithmic curve, with the general form

$$y = k(\ln x) + c,$$

where k and c are constants. Adapting our variables and using

t = number of years after 1940,

$CG(t)$ = number of hectares ($\times 10^6$) cleared for cattle grazing in year $1940 + t$, we have

$$CG(t) = k(\ln t) + c.$$

To use this type of equation as a model, we need to substitute two points from the data provided in order to determine the constants (two constants, two points). Here's an example.

Use the years 1960 and 1980; the points are (20, 0.30) and (40, 1.36). Then substitute into the general form to get the two equations

$$0.30 = k(\ln 20) + c$$

$$1.36 = k(\ln 40) + c,$$

or

$$0.30 = 2.996k + c$$

$$1.36 = 3.689k + c.$$

This is a system of 2 equations in 2 unknowns with 2×2 coefficient matrix

$$A = \begin{bmatrix} 2.996 & 1 \\ 3.689 & 1 \end{bmatrix},$$

and 2×1 constant matrix

$$\begin{bmatrix} 0.30 \\ 1.36 \end{bmatrix}.$$

Enter these matrices into your calculator, and determine the product $A^{-1}B$ to get the solution

$$A^{-1}B = \begin{bmatrix} 1.53 \\ -4.284 \end{bmatrix},$$

so

$$k = 1.53$$
$$c = -4.284.$$

The equation is

$$CG(t) = 1.53\ (\ln t) - 4.284.$$

Differences and error are computed as usual. First evaluate the function at the t-values:

$$t = 20,\ \ CG(20) = 1.53(\ln 20) - 4.284 = 0.299$$
$$t = 30,\ \ CG(30) = 1.53(\ln 30) - 4.284 = 0.920$$
$$t = 40,\ \ CG(40) = 1.53(\ln 40) - 4.284 = 1.360$$
$$t = 48,\ \ CG(48) = 1.53(\ln 48) - 4.284 = 1.639$$

Next compute the differences and sum them to obtain the error:

Year	Predicted	Actual	Difference
1960	0.299	0.300	0.001
1970	0.920	0.950	0.030
1980	1.360	1.360	0.000
1988	1.639	1.400	0.239
			Error = 0.270

And finally, the graph is shown in Figure 11.2.

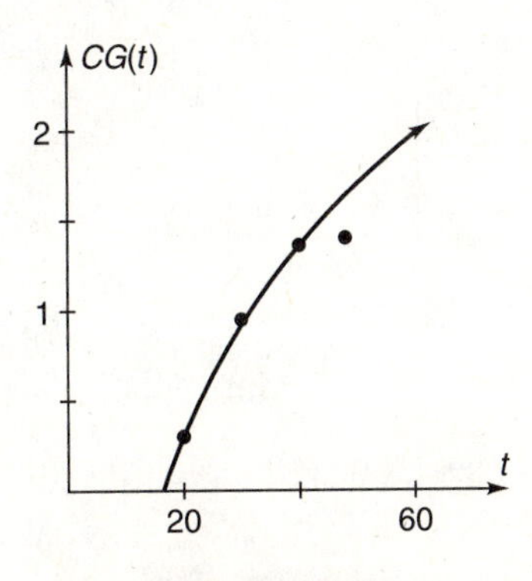

Figure 11.2

Although this may not be the best function to model our data, we use it to illustrate some predictions with a logarithmic function. For example, in the year 2001 ($t = 61$) we predict that the amount of rain forest to be cleared for cattle grazing will be

$$\begin{aligned} CG(61) &= 1.53 \ln (61) - 4.284 \\ &= 2.0, \end{aligned}$$

so in 2001, 2 million hectares will be cleared.

Next, we predict the year in which 2.5 million hectares will be cleared. This involves finding a solution to the equation

$$2.5 = 1.53 \ln t - 4.284.$$

We solve this algebraically by first solving for $\ln t$:

$$1.53 \ln t = 6.784$$

$$\ln t = \frac{6.784}{1.53} = 4.434,$$

then $e^{\ln t} = e^{4.434}$, and $t = 84$ years after 1940, or in the year 2024. Also, this solution can be estimated by using the calculator to find a point on the graph with second coordinate near 2.5. Remember to use "trace." You should see a point near (84.3, 2.50); this means $t = 84$ (rounded) when $CG = 2.50$. Hence the solution is approximately $t = 84$, and the year would be $1940 + 84 = 2024$. See Figure 11.3.

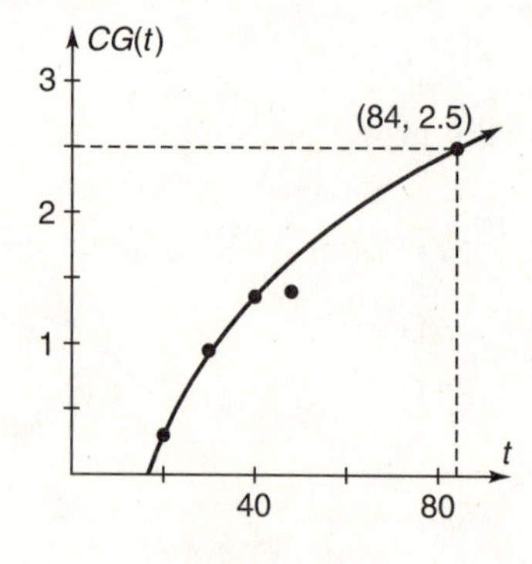

Figure 11.3

Now you have seen one example of writing a logarithmic equation to match the data provided. Not bad. It's almost as easy as writing a linear equation, only now you are using logarithms. They are nice functions. Make friends with them.

group work

The groups should now find the remaining logarithmic equations for this data and make their reports. There are only six equations total, and we've already done one of them. Call the best function $CG(t)$, and save it. Use $CG(t)$ to determine this information:

1. In what year will the amount of rain forest cut for cattle ranching equal 3 million hectares? Solve algebraically and check your answer by estimation on the calculator.
2. How many hectares will be cleared for cattle ranching in the year 2005?

11.3 PREDICTING DEFORESTATION DUE TO AGRICULTURE AND DEVELOPMENT

Although agriculture has been encouraged locally, rain forest soils are so lacking in nutrients necessary for crop growth, it is proving to be too inefficient to pursue. Farmers would abandon their small farms after the few existing nutrients were used up, leaving the land fallow and useless. It takes up to 30 years for the land to regain enough nutrients to farm again, and then only for a short time.

Development plans in most tropical countries include roads, dams, and resettlement of peasant populations. Many of these projects have been, at best, minimally successful, and the damage to the rain forests has proved to be more detrimental than beneficial.

The following data (Chart 11.3) are the estimated amount of rain forest destroyed for agricultural and developmental projects by year. This chart provides the information for the third and final factor which we will use in our analysis of deforestation.

Chart 11.3 Rain Forest Destroyed for Agriculture and Development

Year	Hectares $\times 10^6$
1960	2.21
1970	3.79
1980	4.92
1988	5.77

Groups should now model this information, and report. We *strongly* suggest a logarithmic model here, for the same reasons as before. Use $t = 0$ in 1940, and name your best equation $AD(t)$. Use $AD(t)$ to determine the following: in what year will the amount cut for agriculture and development be five times the amount destroyed for cattle in 1992? Solve algebraically, and check your result by estimation with the calculator.

You have now completed your model for the three major factors involved in deforestation. In the next chapter, we build models which define total emission of carbon from each of the three sources: automobiles, energy consumption, and deforestation.

CHAPTER 12

Total Carbon Dioxide Emission Functions

O.K., now you've got the model for each factor from the three sources of carbon dioxide emission: cars, power, and deforestation. Each group should now concentrate on one of the sources, and write a function which gives total CO_2 or carbon emission from that source as a function of time t. We break this chapter down into three sections. Each group should read the section appropriate for its assigned (or chosen) source.

12.1 PASSENGER CARS

The three factors involved here are: total number of passenger cars, $A(t)$; average fuel efficiency, $MPG(t)$; and average miles per car per year, $M(t)$. The conversion necessary to obtain CO_2 emission is this: each gallon of gasoline burned emits 20 lbs of carbon dioxide (on the average).

In order to determine the function which defines CO_2 emission from automobiles, you must first determine the total number of gallons of gasoline actually burned. See Chart 12.1 for help. Fill in all the blanks, and also the data for 1992 using your derived equations. This should aid you in combining all of your equations to obtain the desired function. You should use only the part of the piecewise function $M(t)$ for which $t \geq 30$ since we will use the total emission function for future predictions. Name this function $CO_2A(t)$, and clearly explain your variables. Note that the domain of $CO_2A(t)$ is $t \geq 30$. Save this important function.

Enter $CO_2A(t)$ into your calculator so that you can easily evaluate and graph it. Maybe you can enter the equation for each factor and then combine them. Before you graph this function, you'll need to set the calculator ranges so that you can see the relevant parts of the graph. For example, if you want to see trends through the year 2050, then you should set x min at 30, and x max at 110 $(2050-110=1940)$ or more. To estimate your vertical range, calculate $CO_2A(30)$ and set y min a little less than this; then calculate $CO_2A(110)$ and set y max a little more than this number.

Chart 12.1 Basic data obtained from the *Statistical Abstracts of the United States*

(Students are to fill in the unknown quantities.)

Year	No. of Passenger Cars $\times 10^6$	Avg. MPG per Car	Avg. Miles per Yr. per Car $\times 10^3$	Total No. of Miles Driven	Number of Gallons of Gas Used	Total CO_2
1940	27.5	14.8	9.1	?	?	?
1950	40.3	13.9	9.1	?	?	?
1960	61.7	13.4	9.5	?	?	?
1970	89.3	13.5	10.0	?	?	?
1980	121.6	15.5	8.8	?	?	?
1986	135.4	18.3	9.3	?	?	?
1992	?	?	?	?	?	?

Suppose you wanted to know the year in which CO_2 emission from passenger cars in the United states will reach the astronomical figure of 2000×10^9 lbs. This involves solving the equation

$$CO_2A(t) = 2000,$$

which is quite difficult. But you can estimate the solution by locating the point on the graph with vertical coordinate approximately 2000. Do it.

Or, maybe you'd need to know CO_2 emission from all these cars when you're sixty-four (remember the Beatles?). To do this, you need to know how old you are now. Figure this out.

See what other interesting things you can predict or otherwise glean from your function. For example, do you see any high or low points on your graph? If so, explain these. Include all of this in your group report for the class.

12.2 POWER CONSUMPTION

You have modeled the amount of power consumed from three sources: coal, natural gas, and petroleum. These equations predict how many quadrillion BTU of each fuel will be consumed in any particular year. One reason for wanting to predict how much energy will be consumed in the United States is so that we can predict how much carbon dioxide will be emitted from these sources. The three fuels whose consumption you have modeled are called carbon-bearing fuels, because when burned each of them emits carbon in the form of carbon dioxide. We must determine how much carbon is due to each of these sources. You have the model for each source: coal consumption, $CC(t)$; petroleum consumption, $PC(t)$; natural gas consumption, $NGC(t)$.

To determine the carbon emission from each of these, you need to know how much carbon is emitted per quad for each fuel. Chart 12.2 below gives the necessary information. In order to keep our numbers relatively small, we choose "gigatons of carbon per quad" to measure emission; one gigaton is 10^{12} kilograms, and recall that one quad is 10^{15} BTU. (See Appendix B for all units.)

Chart 12.2 Carbon Content of Fuels

Fuel	Gigatons per Quad
Coal	.02500
Natural Gas	.01454
Petroleum	.02045

We wish to know how much carbon is emitted from coal, natural gas, and petroleum. For example, to obtain total carbon emitted from the burning of coal in year $1970 + t$, multiply $CC(t)$ by the factor .025.

Power groups should now determine carbon emission from natural gas and petroleum, and then combine all three sources to obtain a total carbon emission function for energy consumption. (Remember that these are piecewise functions; since we will use these to predict future events, you should use the latter definition, i.e., $t \geq 10$.) Name it $TCE(t)$; this defines total carbon emission from energy consumption in gigatons in year

$$1970 + t, t \geq 10.$$

Enter $TCE(t)$ into your calculator so that you can easily evaluate and graph it. Maybe you can enter the equation for each factor and then combine them. Before you graph this function, you'll need to set the calculator ranges so that you can see the relevant parts of the graph. For example, if you want to see trends through the year 2050, you should set x min at 10 and x max at 80 $(2050 - 80 = 1970)$ or more. To estimate your vertical range, set y min a little less than the minimum value of the function; then calculate $TCE(80)$ and set y max a little more than this number.

Suppose you need to know the year in which carbon emission from energy consumption will reach 10 gigatons. This involves solving the equation $TCE(t) = 10$ for t, which is easy with the calculator. Estimate the solution by locating the point on the graph with second coordinate approximately equal to 10. What is your estimate for the first coordinate?

Or, maybe you would like to know carbon emission from energy consumption when you're sixty-four (remember the Beatles?). To do this, you need to know how old you are now.

See what other interesting things you can predict or otherwise glean from your function. For example, do you see any high or low points on your graph? If so, explain these. Include all of this in your group report for the class.

12.3 DEFORESTATION

The three major reasons for deforestation are logging, cattle grazing, and agriculture and development. You have derived the model for each of these factors; retrieve them. To obtain carbon emission, the key conversion is: one hectare of destroyed rain forest emits one half of a metric ton of carbon.

Rain forest groups should now combine the three models for the factors to determine a carbon emission function for deforestation. Name it $CDF(t)$; this defines total carbon emission in metric tons from deforestation in year $1940 + t$.

Enter $CDF(t)$ into your calculator so that you can easily evaluate and graph it. Maybe you can enter the equation for each factor and then combine them. Before you graph this function, you'll need to set the calculator ranges so that you can see the relevant parts of the graph. For example, if you want to see trends through the year 2050, then you should set x min at 0, and x max at 110 $(2050 - 110 = 1940)$ or more.

Estimation of the vertical range is a little tricky because there are logarithms in your function. This means that $CDF(0)$ is undefined (the domain of the natural logarithmic function consists only of positive numbers), so you should calculate *CDF* at some small positive number, say *CDF*(5) and set *y* min to this number or a little less. Then calculate *CDF*(110) and set *y* max a little more than this.

Suppose you wanted to know the year in which carbon emission from deforestation in the United States will reach 10×10^6 tons. This involves solving the equation $CDF(t) = 10$, which is quite difficult. But you can estimate the solution by locating the point on the graph with second coordinate approximately 10. What is your estimation?

Or, maybe you need to know carbon emission from deforestation when you're sixty-four (remember the Beatles?). To do this, you need to know how old you are now.

See what other interesting things you can predict or otherwise glean from your function. For example, do you see any high or low points on your graph? If so, explain these. Include all of this in your group report for the class.

PART III

Accumulation of Carbon Dioxide

INTRODUCTION

In Part III, our primary objective is to study the accumulation of carbon emission due to human activity and how this affects the atmospheric concentration of CO_2. In order to carry out this study, in Chapter 14 we first derive a new model for long-term atmospheric CO_2 concentration using a more sophisticated function, namely an exponential function. This easily enables us to determine the growth rate of atmospheric CO_2. Then we use the most recent information on annual carbon emission and its growth rate to predict future carbon emission and its contribution to atmospheric accumulation.

In Chapter 15, we approximate the models developed in earlier chapters with exponential models, and use these new models to estimate atmospheric CO_2 accumulation due to each of the three sources studied.

The mathematical concepts needed to conduct these studies are linear and exponential functions and geometric series.

CHAPTER 13

Geometric Series

13.1 NOTATION, DEFINITIONS, AND SUMS

A finite *geometric progression* is a finite sequence of numbers, each of which is obtained from its predecessor by multiplication by a fixed constant. The general form for a finite geometric progression is

$$a, ar, ar^2, ar^3, \ldots, ar^n,$$

where a can be any number, the fixed constant is r, and n can be any positive integer. The number r is called the *common ratio*. An example of a particular geometric progression is

$$3, 6, 12, 24, 48.$$

Here $a = 3$, $r = 2$, and $n = 4$.

A finite geometric series, or simply *geometric series,* is

$$a + ar + ar^2 + \ldots + ar^n;$$

i.e., the sum of the terms of a finite geometric progression. The *sum*

$$S = a + ar + ar^2 + \ldots + ar^n$$

of a geometric series is easily determined by the formula

$$S = \frac{a(r^{n+1} - 1)}{r - 1}$$

Note: Chapter 13 is a prerequisite for Chapters 14 and 15.

The following examples illustrate how this formula works.

EXAMPLE 13.1

The geometric series

$$3 + 6 + 12 + 24 + 48$$

corresponds to the progression given as your earlier example. Recall, or determine

$$a = 3$$

$$r = 2$$

$$n = 4.$$

Substitute into the formula to get

$$S = \frac{3(2^5 - 1)}{2 - 1} = 93.$$

Thus

$$93 = 3 + 6 + 12 + 24 + 48.$$

EXAMPLE 13.2

Consider the geometric series

$$5 + 15 + 45 + 135 + 405 + 1215 + 3645 + 10{,}395.$$

First, determine

$$a = 5$$

$$r = 3$$

$$n = 7.$$

So

$$S = \frac{5(3^8 - 1)}{3 - 1} = 16{,}400.$$

EXAMPLE 13.3

Consider the series

$$5.31 + 5.31(.26) + 5.31(.26)^2 + 5.31(.26)^3.$$

In this form it is easy to see that

$$a = 5.31$$

$$r = .26$$

$$n = 3,$$

so

$$S = \frac{5.31((.26)^4 - 1)}{.26 - 1} = 7.143.$$

This last example (13.3) is given in the form

$$a + ar + ar^2 + \ldots + ar^n$$

because the geometric series you will encounter in *Earth Algebra* will come to you in this form. Note that this makes it very easy to determine a, r, and n.

You can also write your own geometric series if you know the numbers a, r, and n. The next example illustrates.

EXAMPLE 13.4

Let $a = 7$, $r = 4$, and $n = 6$.

The corresponding series is

$$7 + 7(4) + 7(4)^2 + 7(4)^3 + 7(4)^4 + 7(4)^5 + 7(4)^6,$$

or multiplied out,

$$7 + 28 + 112 + 448 + 1{,}792 + 7{,}168 + 28{,}672.$$

Its sum is

$$S = \frac{7(4^7 - 1)}{4 - 1} = 38{,}227.$$

13.2 THINGS TO DO

For 1–5, determine the numbers a, r, and n, and the sum of each geometric series.

1. $1 + 6 + 36 + 216 + 1{,}296 + 7{,}776 + 46{,}656 + 279{,}936$.
2. $8 + 4 + 2 + 1 + .5 + .25 + .125 + .0625$.
3. $71.2 + 71.2(4.3) + 71.2(4.3)^2 + 71.2(4.3)^3 + \dots + 71.2(4.3)^{12}$.
4. $1.6 + 1.6(.431) + 1.6(.431)^2 + 1.6(.431)^3 + 71.2(.431)^4$.
5. $.5 + .5(.1) + .5(.1)^2 + \dots + .5(.1)^{793}$.

For 6–8, write the geometric series corresponding to the given values of a, r, and n, and find its sum.

6. $a = 3, r = 8, n = 5$.
7. $a = 4.1, r = 9.6, n = 7$.
8. $a = 2.16, r = 1.7, n = 52$.

CHAPTER 14

Carbon Dioxide Concentration Revisited

14.1 LONG-TERM CARBON DIOXIDE CONCENTRATION

When we first began modeling, we modeled atmospheric concentration of carbon dioxide with a linear function, which wasn't a bad first try. Let's now look at more long-term data and try a different model which will better reflect these trends. The measurements are made at the Mauna Loa Observatory in Hawaii. (See Chart 14.1)

Chart 14.1

Year	Average CO_2 Concentration in ppm
1900	280.0
1950	310.0
1975	331.0
1980	338.5
1988	351.3

We plot the corresponding points on a $(t,\ CO_2)$ coordinate system (Figure 14.1) where

$$CO_2(t) = \text{carbon dioxide concentration in ppm in year } 1900 + t.$$

(Note: for this model $t = 0$ in 1900, so t = number of years after 1900.) This graph is shown in Figure 14.1.

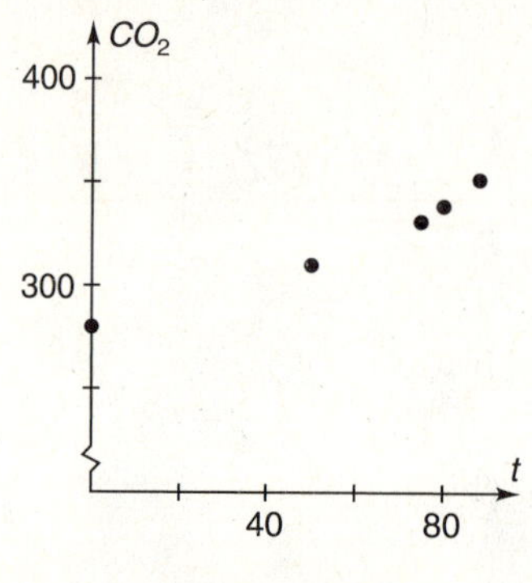

Figure 14.1

Next, we compute the slopes of the line segments which join adjacent points. The slope between the first two points, (0, 280) and (50, 310), is

$$m = \frac{310 - 280}{50 - 0} = .6$$

and the slope between (50, 310) and (75, 331) is

$$m = \frac{331 - 310}{25} = .84.$$

The next two slopes, in order, are 1.5 and 1.8. These slopes are increasing, so either a parabolic or exponential model will provide a better fit than a linear one. Since there is no indication of a vertex in our data, we use an exponential model.

Recall the general form for an exponential function

$$f(x) = a(b^x),$$

where a and b are constant (Chapter 10). Adapt this form to our notation for this data,

$$CO_2(t) = a(b^t).$$

As an example, we derive the exponential equation whose graph passes through the points (50, 310.0) and (75, 331.0) which correspond to data in 1950 and 1975. Substitution gives the two equations

$$310 = a(b^{50})$$

$$331 = a(b^{75}).$$

which we must solve for a and b. Divide the second equation by the first,

$$\frac{331}{310} = \frac{a(b^{75})}{a(b^{50})},$$

which eliminates a, and yields

$$b^{25} = 1.0677.$$

Solve this equation on your calculator to obtain

$$b = 1.0026.$$

Now substitute this value for b into the first equation and get

$$310 = a(1.0026^{50}),$$

and hence

$$a = 272.26.$$

The exponential equation is

$$CO_2(t) = 272.26(1.0026)^t,$$

where $CO_2(t)$ = atmospheric concentration in ppm in year 1900 + t. Chart 14.2 shows differences and error for this function.

Chart 14.2

t	$CO_2(t)$	Actual CO_2 Concentration	Difference
0	272.3	280.0	7.7
50	310.0	310.0	0.0
75	330.8	331.0	0.2
80	335.1	338.5	3.4
88	342.2	351.3	9.1

Error: 20.4

Figure 14.2 shows the graph of this function with the original data points in their correct relative positions.

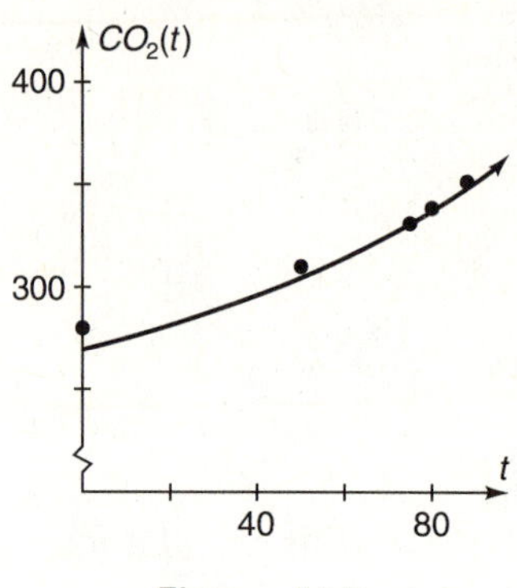

Figure 14.2

Carbon dioxide concentration is increasing as evidenced by the data, and of course this is reflected in the exponential equation

$$CO_2(t) = 272.3(1.0026)^t.$$

In fact in any one year period, the amount of increase is

$$CO_2(t+1) - CO_2(t) = 272.3\,(1.0026^{t+1}) - 272.3\,(1.0026^t).$$

Factor out $272.3(1.0026^t)$ from this expression to get

$$272.3(1.0026)^t\,[1.0026 - 1] = 272.3(1.0026^t)\,(.0026).$$

This last expression can be used to compute the amount of increase in carbon dioxide concentration in any year $1900 + t$, and we leave it in this unsimplified form for a reason.

Next, we can see the percentage change by dividing the annual increase by the total concentration in year $1900 + t$ to get

$$\frac{272.3(1.0026^t)(.0026)}{272.3(1.0026^t)} = .0026.$$

This number, .0026, says that the CO_2 concentration has increased by 0.26% in year $1900 + t$. Note that this is totally independent of t; i.e., it is the same regardless of year. This number 0.26% is known as the *growth rate*. (*Growth rate* is the annual percentage change.) The growth rate can be easily determined in this manner for any exponential function.

Determine the remaining nine exponential equations using the CO_2 concentration data in Chart 14.1. Determine the error for each and find the best function.

Use the best equation to determine each of the following:

1. CO_2 concentration in 2010;
2. the year when concentration will be 560 ppm, which is double the preindustrial level;
3. the growth rate. (Can you find a general formula for growth rate?)

In Chapter 3, you modeled U.S. population with a linear equation. Now, we look at world population, which requires a more sophisticated model. The chart below shows population in billions worldwide in the indicated year. Model these data with an exponential equation using $t = 0$ in 1950; then use your model to answer the questions below.

Year	World Population ($\times 10^9$)
1950	2.6
1960	3.1
1970	3.7
1980	4.5
1989	5.3

1. Predict world population in the year 2020.
2. When will world population double its present size?
3. Determine the growth rate of world population.
4. Suppose that, as of this year, this growth rate is decreased by 10%. How would this effect the population in the year 2020?

14.2 CARBON EMISSION AND INCREASE IN CO_2 CONCENTRATION

When carbon is released into the air as carbon dioxide, some of it is reabsorbed by plants and the oceans as part of the carbon cycle, something like the way blood circulates. A portion of the carbon dioxide that is released into the air by human activities stays in the atmosphere. We can estimate the effect this has on climate if we can somehow describe how carbon dioxide emitted from human sources accumulates in the atmosphere and how this affects the concentration of CO_2. We discuss this problem in terms of carbon instead of carbon dioxide since most available information is given this way. If we know the rate of growth of carbon emission (either from all human activity or from some particular source), it is not difficult to estimate how much that will raise the carbon dioxide concentration in the atmosphere. The study below predicts the worldwide carbon emissions by year beginning in 1991.

It was estimated that the amount of carbon in 1991 in the atmosphere was 750 gigatons. A gigaton, abbreviated GT, is a billion metric tons (10^9 metric tons), and is a convenient unit to use. Also, in 1991 the annual emission was about 7 GT. In the last few years, the amount of carbon being emitted annually has been increasing at a rate of about 1% per year but during most of the eighties the amount of carbon being emitted grew at a rate more like 1.7%. A rate of 1% means that if 7 gigatons of carbon are emitted this year, then next year, the emissions will be 7.07 GT (the 1991 level plus an additional 1% of the 1991 level:

$$7 + 7(.01) = 7(1 + .01) = 7(1.01).$$

In the year following, another 1% increase yields

$$\begin{aligned} 7(1.01) + (.01)[7(1.01)] &= 7(1.01)(1 + .01) \\ &= 7(1.01)(1.01) \\ &= 7(1.01)^2 \end{aligned}$$

GT of carbon emitted. In general, if t is the number of years after 1991, the emission will be

$$7(1.01)^t \text{ GT carbon.}$$

Under the assumption that the 1991 carbon emission was 7 GT and that the growth rate continues to be 1%, we determine the increase in atmospheric carbon dioxide concentration due to worldwide carbon emission.

The equation which can be used to predict the amount of carbon emitted annually is

$$C(t) = 7(1.01)^t,$$

where $C(t)$ = gigatons of carbon emitted worldwide in year 1991 + t. This function $C(t)$ describes annual emission. Next, we write a function which defines total emission for the period from 1991 through 1991 + t. This function will be the sum

$$7 + 7(1.01) + 7(1.01)^2 + \ldots + 7(1.01)^t,$$

which is a geometric series with $a = 7$, $r = 1.01$, and $n = t$. Its sum is

$$\frac{7\left[(1.01)^{t+1} - 1\right]}{1.01 - 1}$$

This is the total amount of carbon emitted worldwide over this period. Although not all carbon emitted remains in the atmosphere, 58% of it does. (This estimate, .58, is known as the airborne fraction.) Hence the total carbon (in GT) emitted from 1991 through 1991+ t which remains in the atmosphere is

$$\frac{(.58)(7)\left[(1.01)^{t+1} - 1\right]}{1.01 - 1}.$$

Note that this expression only gives carbon accumulation from 1991 through 1991 + t, but total atmospheric carbon can be obtained by adding to this quantity the estimated 1991 amount of carbon in the atmosphere, which is 750 gigatons. This yields the function

$$AC(t) = 750 + \frac{(.58)(7)\left[(1.01)^{t+1} - 1\right]}{1.01 - 1} = 750 + 406\left[(1.01)^{t+1} - 1\right].$$

where $AC(t)$ = total atmospheric carbon in gigatons in 1991 + t.

This function can be used to predict the total atmospheric carbon in any year after 1991. For example, in year 2010,

$$t = 2010 - 1991 = 19,$$

so total atmospheric carbon will be

$$AC(19) = 839.4 \text{ gigatons.}$$

Derive a similar function which defines total atmospheric carbon if the growth rate were reduced to .5%. Assume other data remain the same.

What we really want to know is what this accumulation of carbon means in terms of atmospheric concentration of CO_2. So we need to write an equation which defines *ppm* as a function of *AC*. We assume this equation to be linear, and specify two points which will determine this equation.

In 1991 the total carbon in the atmosphere was 750 GT and the concentration was 353 ppm. Certainly if there were no carbon, then the concentration would be 0 ppm. Now find the equation of the line through the points (0, 0) and (750, 353). The line has slope .471, and with variables (*AC*, *ppm*), the equation is

$$ppm(AC) = .471AC.$$

Now put the pieces together and determine the concentration as a function of time t. That is just the composite function:

$$\begin{aligned} ppm(AC(t)) &= .471[750 + 406(1.01^{t+1} - 1)] \\ &= 353.25 + 191.226(1.01^{t+1} - 1). \end{aligned}$$

Using this new model, we determine when atmospheric concentration will be twice the preindustrial level of 280 ppm. (Remember that this is when the temperature will increase about 5.4°F.) This can be done by solving the equation

$$560 = 353.25 + 191.226(1.01^{t+1} - 1).$$

Using the trace, we determine

$$t = 72.7.$$

This predicts that the CO_2 concentration will double 73 years after 1991, or in the year 2064. Note that this is much earlier than our previous models predict; this is due to recent increased carbon emission.

Use the $AC(t)$ function with the .5% growth rate which you derived in the previous exercise to determine the year in which this doubling of preindustrial CO_2 will occur under this new assumption.

CHAPTER 15

Contribution to Atmospheric CO_2 Concentration from Three Sources

In this chapter we estimate the amount of carbon dioxide which will be added to the atmosphere from each of the three sources over the next ten years. One way to accomplish this is to evaluate each respective total emission function at t values corresponding to each year from 1992 through 2002, sum the answers, and multiply by the airborne fraction. This is a tedious method, particularly if a twenty- or thirty-year period were of interest. So, we will derive a new function which makes these computations quicker and easier.

Let $F(t)$ denote the total emission function for one of the three sources (if automobiles, $F(t) = CO_2A(t)$; if energy, $F(t) = TCE(t)$; if deforestation, $F(t) = CDF(t)$). The new function will be exponential, and will closely approximate $F(t)$. This you can see on your calculator when working through your respective section in this chapter. There are three advantages to using an exponential function:

1. the approximation to $F(t)$ is good;
2. a growth rate can easily be determined;
3. it is easy to sum the annual accumulations.

To derive the exponential model we need two points: compute $F(t)$ for values of t corresponding to the years 1992 and 2002. This gives the points which will be used to derive the exponential function $E(t)$. You may think we are being inconsistent because we are only using two particular points to model emission. We are not finding all equations using all pairs of points and computing errors to find the best. There is a difference in this case; previously we have been modeling *real* data, and here we are approximating *predicted* data points.

Each group should now complete this study by following the steps listed below. Each group should do the study for its chosen or assigned CO_2 source. Recall,

$$F(t) = \text{original derived total emission function,}$$

and

$$E(t) = \text{exponential approximation to } F(t)\,,$$

as described above.

Each group should conduct the study below for its assigned source, enter automobiles, energy, or deforestation.

1. Derive the exponential function $E(t)$.
2. Simultaneously graph $F(t)$ and $E(t)$ on your calculator, with x-range set at x min corresponding to the t value for 1992, and x max corresponding to the t value for 2002. Here you should see the close approximation of $F(t)$ by $E(t)$.
3. Adjust the function $E(t)$ to get new a exponential function $E_0(t) = a_0 b^t$, where t now equals 0 in 1992. This requires a substitution for t and simplification to obtain the form $a_0 b^t$. (In this step, t has been translated so that $t = 0$ in the first year of the period of interest.)
4. Determine the growth rate for your source of CO_2 or carbon emission. Recall, growth rate is annual percentage change, and hence is

$$\frac{a_0 b^{t+1} - a_0 b^t}{a_0 b^t} = \frac{a_0 b^t (b-1)}{a_0 b^t} = \mathrm{b} - 1.$$

5. Compute the total CO_2 emission from your source during the time period 1992 – 2002. This is the sum

$$E_0(0) + E_0(1) + \ldots + E_0(10),$$

which is a geometric series (see Chapter 13).

6. Ultimately, you want to know the increase in atmospheric concentration due to your source. Units necessary to determine this are gigatons of carbon. Therefore, you may need to convert the total emission from your source from 1992 through 2002 using some of these factors:

   weight CO_2 = weight carbon $\times$ 3.667;

   1 lb. = .4545 kg;

   1 metric ton = 10^3 kg;

   1 gigaton = 10^9 metric tons.

   Determine the amount of carbon from your source which actually remains in the atmosphere (use the airborne fraction). Then use the equation derived in Chapter 14, Section 14.2, to determine the resulting increase in atmospheric concentration (in ppm).

7. Suppose the growth rate of carbon which you computed in Step 4 above was reduced by 10%. This would give a new exponential function $E_1(t)$ which would be of the form $a_0 b_1{}^t$, where

   $$b_1 - 1$$

   is this new growth rate (note a_0 is the same as the a_0 in $E_0(t)$). Repeat Steps 5 and 6 to determine the increase in atmospheric carbon (in ppm) from your source with the reduced growth rate. Compare your two answers.

PART IV

Connecting Carbon Dioxide, People, and Money

INTRODUCTION

From our studies of cars, energy consumption, and deforestation, we have learned that CO_2 emission from each of these sources is sharply increasing; i.e., the atmospheric concentration of carbon dioxide is becoming greater and greater. But, think for a minute: cars don't just drive around by themselves, healthy trees don't just fall down, and most animals don't need electric lights. People drive cars, people cut down rainforests, and people need electricity. In Chapter 17, we use models of U.S. and world population which were dervied earlier to study connections between people and carbon dioxide.

The three sources of atmospheric carbon dioxide which we have studied also happen to be major factors in the U.S. economy, the health of which is sometimes measured by the gross national product, abbreviated GNP. This is the monetary value of all goods produced in the United States during a defined time period. Generally, it is a good measure of the state of the economy, but not always. So it would seem reasonable that there would be a relationship between GNP and things such as number of cars, some portion of deforestation, and energy consumption. In Chapter 18, we model GNP and write certain factors involved in our study of CO_2 emission as functions of GNP. In this way we can study the relationship between GNP and atmospheric CO_2. It kind of gives a relationship between money and global warming.

The new mathematical topic necessary to conduct these studies is the concept of an inverse function. It is covered in Chapter 16.

CHAPTER 16

Inverse Functions

16.1 NOTATION AND DEFINITIONS

Some functions have related functions which are called their *inverses*. *Earth Algebra* functions which have inverses are nonconstant linear, exponential, and logarithmic. If f denotes a function which has an inverse, then its inverse function is denoted by f^{-1}. (This does *not* mean $\frac{1}{f}$.) The inverse of a function "undoes" what the function does.

In general, if $f(x) = y$, then $f^{-1}(y) = x$. This is the defining property of an *inverse function.* This also shows us how to find the inverse for a particular function: if $f(x)$ is defined by the equation

$$y = f(x),$$

we can determine the equation which defines the inverse f^{-1} if we can solve for x in terms of y to get

$$x = f^{-1}(y).$$

Not all functions have inverses. If solving for x in terms of y produced two or more x-values, then the "inverse" would not satisfy the definition of function.

Note: Chapter 16 is a prerequisite for Chapters 17 and 18.

If we form both composites of a function and its inverse, we get

$$f^{-1}[f(x)] = x \text{ and } f[f^{-1}(y)] = y.$$

The next three sections illustrate how to determine the inverse of linear, exponential, and logarithmic functions.

16.2 INVERSES OF LINEAR FUNCTIONS

Two examples illustrate the technique used to determine the inverse of a linear function.

EXAMPLE 16.1

Determine the inverse of $f(x) = 4x + 7$.

To find the inverse of this function, replace $f(x)$ by y to get

$$y = 4x + 7.$$

Then solve for x.

$$y - 7 = 4x,$$

$$x = \frac{y-7}{4}$$

so

$$f^{-1}(y) = \frac{y-7}{4}.$$

We check one of the two composites.

$$f^{-1}[f(x)] = \frac{(4x+7)-7}{4} = x.$$

Similarly, $f[f-1(y)] = y$.

EXAMPLE 16.2

Determine the inverse of $f(x) = 1.37x + 132.5$.

Replace $f(x)$ by y:

$$y = 1.37x + 132.5.$$

Solve for x:

$$1.37x = y - 132.5$$

$$x = \frac{y - 132.5}{1.37}.$$

Hence

$$f^{-1}(y) = \frac{y - 132.5}{1.37}.$$

This time check the other composite, $f[f^{-1}(y)]$:

$$f\left[f^{-1}(y)\right] = 1.37\left(\frac{y - 132.5}{1.37}\right) + 132.5.$$

Similarly, $f^{-1}[f(x)] = x$.

16.3 INVERSES OF EXPONENTIAL FUNCTIONS

The technique for finding the inverse of an exponential function is the same as that for a linear function, but a little more difficult.

You need to remember one important property of the natural logarithmic function (see Chapter 10):

$\ln r^s = s \ln r$, where r is any positive number, and s can be any number.

EXAMPLE 16.3

Determine the inverse of $f(x) = e^x$.

Replace $f(x)$ by y to get

$$y = e^x.$$

Note that you need to solve for x, which is the exponent, and the important logarithmic property enables you to do this. Take the logarithm of each side of $y = e^x$ to get

$$\ln y = \ln (e^x) = x \ln e,$$

and now recall from Section 10.2 that $\ln e = 1$. Hence

$$\ln y = x,$$

or

$$f^{-1}(y) = \ln y.$$

Note that this says the natural logarithmic function is the inverse of the natural exponential function, and therefore in terms of composites,

$$\ln (e^{x}) = x, \text{ and } e^{\ln y} = y.$$

EXAMPLE 16.4

Determine the inverse of $f(x) = e^{2x+3}$. Now

$$y = e^{2x+3}$$

$$\ln y = \ln e^{2x+3}$$

$$\ln y = 2x + 3.$$

Solve for x:

$$\ln y - 3 = 2x$$

$$x = \frac{\ln y - 3}{2}$$

So $f^{-1}(y) = \dfrac{\ln y - 3}{2}$.

To illustrate the relationship between f and its inverse, we observe

$$\begin{aligned} f^{-1}[f(x)] = f^{-1}(e^{2x+3}) &= \frac{\ln (e^{2x+3}) - 3}{2} \\ &= \frac{(2x + 3)\ln e - 3}{2} \\ &= \frac{2x + 3 - 3}{2} \\ &= \frac{2x}{2} = x. \end{aligned}$$

Without computation, do you know what $f^{-1}[f(2)]$ is?

EXAMPLE 16.5

Determine the inverse of $f(x) = 2^x$. Here

$$y = 2^x$$
$$\ln y = \ln 2^x$$
$$\ln y = x \ln 2 = x(.693),$$

and

$$x = \frac{\ln y}{.693},$$

so

$$f^{-1}(y) = \frac{\ln y}{.693}.$$

EXAMPLE 16.6

Determine the inverse of $f(x) = 2.35(1.1^x)$.

Set $y = 2.35(1.1^x)$ and solve for x. First you must divide by 2.35 to get

$$\frac{y}{2.35} = 1.1^x;$$

then take logarithms:

$$\ln\left(\frac{y}{2.35}\right) = \ln(1.1^x)$$
$$= x \ln(1.1)$$
$$= x(.095).$$

This gives

$$x = \frac{1}{.095} \ln\left(\frac{y}{2.35}\right)$$

so

$$f^{-1}(y) = \frac{1}{.095} \ln\left(\frac{y}{2.35}\right).$$

16.4 INVERSES OF LOGARITHMIC FUNCTIONS

The following examples illustrate techniques for finding inverses of natural logarithmic functions. You will need to recall these properties of logarithms:

$$e^{\ln x} = x \text{ and } \ln e^x = x.$$

EXAMPLE 16.7

Determine the inverse of $f(x) = 2 + 3 \ln x$.

Replace $f(x)$ by y to get

$$y = 2 + 3 \ln x.$$

Solve for $\ln x$:

$$\ln x = \frac{y-2}{3};$$

Thus

$$e^{\ln x} = e^{\frac{y-2}{3}},$$

so

$$x = e^{\frac{y-2}{3}},$$

and hence

$$f^{-1}(x) = e^{\frac{x-2}{3}}.$$

We check composites both ways.

$$f[f^{-1}(x)] = 2 + 3 \ln (e^{\frac{x-2}{3}})$$

$$= 2 + 3 \frac{(x-2)}{3}$$

$$= x;$$

and

$$f^{-1}[f(x)] = e^{\frac{(2+3\ln x)-2}{3}} = e^{\ln x}$$

$$= x,$$

just as it should be.

EXAMPLE 16.8

Determine the inverse of $g(x) = 3.7 + 5 \ln (2x)$.

Set

$$y = 3.7 + 5 \ln (2x)$$

and solve for $\ln (2x)$.

$$\ln (2x) = \frac{y - 3.7}{5}$$

so

$$e^{\ln(2x)} = e^{\frac{y-3.7}{5}},$$

$$2x = e^{\frac{y-3.7}{5}}$$

$$x = \frac{1}{2} e^{\frac{y-3.7}{5}}.$$

Thus

$$g^{-1}(y) = \frac{1}{2} e^{\frac{y-3.7}{5}}.$$

Now check the composites:

$$g[g^{-1}(y)] = 3.7 + 5 \ln [2\left(\frac{1}{2}\right) e^{\frac{y-3.7}{5}}]$$

$$= 3.7 + 5 \ln e^{\frac{y-3.7}{5}}$$

$$= 3.7 + 5\left(\frac{y-3.7}{5}\right) = y;$$

and

$$g^{-1}[g(x)] = \frac{1}{2} e^{\frac{[3.7 + 5\ln(2x)] - 3.7}{5}}$$

$$= \frac{1}{2} e^{\ln(2x)}$$

$$= \frac{1}{2}(2x)$$

$$= x.$$

16.5 THINGS TO DO

For 1–5, determine the inverse of the given linear function, then form both composites to show that $f^{-1}[f(x)] = x$ and that $f[f^{-1}(y)] = y$.

1. $f(x) = 2x - 1$.
2. $F(x) = 4 - 3x$.
3. $g(x) = \dfrac{x}{5} + 20$.
4. $h(x) = 7.314x - 2.001$.
5. $f(x) = -5.4x$.

For 6–10, determine the inverse of the given exponential function, form both composites to show that $f^{-1}[f(x)] = x$ and $f[f^{-1}(y)] = y$, and compute the indicated functional values.

6. $g(x) = 3^x$. Compute $g(2)$ and $g^{-1}(9)$.
7. $h(x) = 2.1\,(7.01)^x$. Compute $h(4.3)$ and $h^{-1}(9095.085205)$.
8. $f(x) = (6.03)^x$. Compute $f(2.5)$ and $f^{-1}(2.5)$.
9. $g(x) = -3(1.2)^x$. Compute $g(1)$ and $g^{-1}(-3.6)$.
10. $F(x) = 1 + 5^x$. Compute $F(0)$ and $F^{-1}(2)$.

For 11–15, determine the inverse of the given logarithmic function, form both composites to show that $f^{-1}[f(x)] = x$ and that $f[f^{-1}(y)] = y$, and compute the indicated functional values.

11. $f(x) = 2.7 + 1.3 \ln x$; compute $f(2)$ and $f^{-1}(3.601)$.

12. $g(x) = 5 - 3 \ln x$; compute $g(7)$ and $g^{-1}(-.8377)$.

13. $h(x) = 1.5 \ln (3x) - 9$; compute $h(8)$ and $h^{-1}(-4.2329)$.

14. $L(x) = \ln (5 - x)$; compute $L(0)$ and $L^{-1}(1.6094)$.

15. $U(x) = .874 \ln (2x)$; compute $U(-3)$.

CHAPTER 17

People

17.1 HOW MANY PEOPLE ARE THERE?

In this chapter, we study connections between population and carbon dioxide. First, we derive a relationship between U.S. population and atmospheric carbon dioxide concentration. (Even though U.S. population is only 4.7% of world population, it is responsible for 25% of CO_2 emissions from human activity.)

You derived the U.S. population model in Chapter 3, Section 3.1; that model is

$$P(t) = 2.36t + 179.3, \text{ where}$$

$$P(t) = \text{total U.S. population in millions in } 1960 + t.$$

Also in Chapter 3, the atmospheric CO_2 concentration was derived as the linear model

$$CO_2C(t) = 1.44t + 280, \text{ where}$$

$$CO_2C(t) = \text{atmospheric } \mathrm{CO_2} \text{ concentration in ppm in year } 1939 + t.$$

We wish to study the relationship between U.S. population and CO_2 concentration, so a function which defines CO_2C in terms of U.S. population would be appropriate; i.e., the function we need is $CO_2C(P)$. In order to determine this, we can find the inverse of the population function and form a composite with the CO_2 concentration function.

But first, there is a time discrepancy in the two equations; this needs to be adjusted so that both t's are 0 in the same year. In the population equation, $t=0$ in 1960, whereas in the CO_2C equation, $t=0$ in 1939 (a 21-year difference). We adjust so that $t=0$ in 1939 for both. This involves replacing t by $t-21$ in the $P(t)$ equation to get the new population model (which we still call $P(t)$)

$$P(t) = 2.36(t-21) + 179.3$$

$$P(t) = 2.36t + 129.74,$$

where now

$$P(t) = \text{U.S. population } (\times 10^6) \text{ in } 1939 + t.$$

Next, we find the inverse of this population equation. Replace $P(t)$ by P to get

$$P = 2.36t + 129.74$$

and solve for t:

$$2.36t = P - 129.74$$

$$t = \frac{P - 129.74}{2.36},$$

so

$$t = .424P - 54.975.$$

This is the inverse of the adjusted population equation, which we substitute for t in the CO_2C equation,

$$CO_2C(t) = 1.44t + 280$$

$$CO_2C(P) = 1.44(.424P - 54.975) + 280,$$

so

$$CO_2C(P) = .611P + 200.836.$$

Remember that population P is in millions and CO_2C is in parts per million.

We can use this function $CO_2C(P)$ to answer such interesting questions as: what population increase would correspond to an increase in an average global temperature of 1°F? In order to answer this, we recall our $GT(CO_2C)$ function from Chapter 5 and use it to determine the CO_2C increase which produces a 1°F increase in global temperature. Average temperature increase since the pre-industrial CO_2C level of 280 ppm is defined by the equation:

$$GT(CO_2C) = .0193CO_2C - 5.4.$$

The slope of this line is .0193 which means that a 1 ppm increase in CO_2 concentration corresponds to a .0193°F in average global temperature; the increase in CO_2 concentration which would result in a 1°F increase is

$$\frac{1°F}{.0193°F} = 51.813 \text{ ppm.}$$

The slope of the $CO_2C(P)$ function is .611. This means that an increase of one million people corresponds to an increase in CO_2 concentration of .611 ppm. Hence the population increase which would correspond to a CO_2 concentration increase of 51.813 ppm is

$$\frac{51.813 \text{ ppm}}{.611 \text{ ppm}} = 84.8 \text{ million people.}$$

group work

You can use the $P(t)$ function to predict how many years it will take for the present global temperature to increase 1°F.

Use the long-term exponential model for CO_2 from Section 14.1 and the exponential model for world population, also from Section 14.1, to express CO_2 concentration in terms of world population.

17.2 OF PEOPLE AND CO_2 SOURCES

Each group should read the section appropriate to its assigned or chosen source. In these studies, each group will relate population to one factor involved in CO_2 emission from its source.

Passenger Cars

As long as cars are readily available to us, the more people there are, probably the more cars there will be. So to see that relationship directly, you should study the connection between number of cars and population. To complete this study, follow the steps delineated below.

group work

1. Recall the original U.S. population model (Section 3.1)

$$P(t) = 2.36t + 179.3,$$

where $t = 0$ in 1960, and recall the $A(t)$ model for number of U.S. automobiles (Section 8.1). Adjust the above $P(t)$ function to get a new $P(t)$ function in which $t = 0$ in 1940 (base year for autos).

2. Find the inverse of the adjusted $P(t)$ function.
3. Form the composite function $A(P)$ which defines number of cars in terms of U.S. population.
4. Provide a verbal interpretation of the slope of $A(P)$.
5. Determine the present U.S. population, and the projected U.S. population ten years from now.
6. Determine CO_2 emission from automobiles this year and the projected emission ten years from now (use $CO_2A(t)$ from Chapter 12).
7. Suppose the rate of population growth slows by 10%, i.e., from the current rate of 2.36×10^6 to 2.12×10^6 per year. What would be the number of automobiles corresponding to this decrease ten years from now? What effect would this have on CO_2 emission ten years from now?

Energy Consumption

As the population grows, more and more electricity will be used. A major portion of the electricity in the United States is generated from coal burning power plants, so it seems reasonable to study the relationship between the population and coal consumption. To do this study, follow the steps listed below.

1. Recall the original coal consumption function $CC(t)$ (Section 9.1). This is a piecewise function with the first part parabolic and the second part linear. The second part corresponds to the most recent years, and since we will be concentrating on recent and future time periods in this study, we only need consider that part of this piecewise function,

   $$CC(t) = .419t + 11.243,$$

   where $t = 0$ in 1970, but $t \geq 10$.

   Now, also recall the original U.S. population function (Section 3.1),

   $$P(t) = 2.36t + 179.3,$$

   where $t = 0$ in 1960. Adjust this function to get a new population function, also called $P(t)$, where $t = 0$ in 1970. This is necessary so that the t in both $CC(t)$ and $P(t)$ coincide.
2. Find the inverse of the adjusted $P(t)$ function.
3. Form the composite function $CC(P)$, which defines coal consumption in the U.S. in terms of U.S. population.
4. Write a verbal interpretation of the slope of $CC(P)$.
5. Determine the present U.S. population and the projected U.S. population ten years from now.
6. Determine carbon emission from energy consumption this year and the projected emission ten years from now (use $TCE(t)$ from Chapter 12).
7. Suppose the rate of population growth slows by 10%, i.e., from the current rate of 2.36×10^6 to 2.12×10^6 per year. What would be the corresponding coal consumption with this decrease in ten years? What effect would this have on carbon emission from energy consumption in the U.S. ten years from now?

Deforestation

As the population increases, so does the demand for goods, in particular for furniture, homes, and many other wood products. The United States is an importer of wood cut from rain forest, so a study of the relationship between population and deforestation for logging seems reasonable. To do this study, follow the steps outlined in the following box.

1. Recall the original U.S. population function (Section 3.1)

   $P(t) = 2.36t + 179.3,$

   where $t = 0$ in 1960, and recall the logging function $L(t)$ from Section 11.1. Adjust the above function $P(t)$ to get a new population function $P(t)$ in which $t = 0$ in 1940 (base year for deforestation). This is necessary so that the t in both $P(t)$ and $L(t)$ coincide.
2. Find the inverse of the adjusted $P(t)$ function.
3. Form the composite function $L(P)$ which defines hectares of rain forest cut for logging in terms of population of the United States.
4. The function $L(P)$ is, of course, not linear so its slope is not defined. You can, however, use the slope concept to approximate the rate of increase of logging with respect to population. Here's how: determine the U.S. population P_0 this year and evaluate $L(P)$ for this number, i.e., $L(P_0)$. Then evaluate $L(P)$ for this year's population plus one million, i.e., $L(P_0 + 1)$. The difference $L(P_0 + 1) - L(P_0)$ is a "slope approximation." (Recall that if $y = f(x)$ is a linear function, then if x increases by 1 unit, the change in y is the slope. Also recall that the population P is given in millions, so increasing the population by one million means $P + 1$.) Now, write a verbal interpretation of the current rate of increase of logging in the rain forest with respect to population.
5. Determine the present U.S. population and the projected population ten years from now.
6. Determine carbon emission due to deforestation this year and the projected emission ten years from now (use $CDF(t)$ from Chapter 12).

7. Suppose the rate of growth of the U.S. population slows by 10%; i.e., from the current rate of 2.36×10^6 to 2.12×10^6 per year. What would be the number of hectares cut for logging corresponding to this decrease ten years from now? What effect would this have on carbon emission due to deforestation ten years from now?

CHAPTER 18

Money

18.1 MODELING GROSS NATIONAL PRODUCT

In this section, we will model GNP and write certain factors involved in our study of CO_2 emission as functions of GNP. This way we can study the relationship between GNP and CO_2 in the atmosphere. It kind of gives a relationship between money and global warming.

The information in chart 18.1 once again comes from the *Statistical Abstract of the United States*. Figure 18.1 is a (t, GNP) coordinate system with corresponding points plotted.

Chart 18.1 Gross National Product

Year	(Constant 1982 dollars $\times 10^{12}$)
1960	1.665
1970	2.416
1980	3.187
1989	4.118

1. Plot the corresponding points on a (t, GNP) coordinate system; then model this data using an exponential equation. Name your best function $GNP(t)$, where $t = 0$ in 1960.

2. Recall the total CO_2 concentration model from Chapter 14,

$$CO_2(t) = 280\,(1.0024^t),$$

where $t = 0$ in 1900. Adjust the function $CO_2(t)$ to obtain a new $CO_2(t)$ function where $t = 0$ in 1960.

3. Find the inverse of the $GNP(t)$ function from exercise 1.

4. Form the composite function $CO_2(GNP)$, which defines CO_2 as a function of GNP.

5. Graph the function $CO_2(GNP)$.

6. The function $CO_2(GNP)$ is not linear, so its slope is not defined. However, you can use the concept of the slope to approximate the rate of increase of CO_2 with respect to GNP. (Some of you may have already done something like this in Section 17.2.) Here's how: determine the gross national product GNP_0 this year and evaluate $CO_2(GNP)$ for this number; i.e., $CO_2(GNP_0)$. Then evaluate $CO_2(GNP)$ for this year's gross national product plus 1 billion dollars, i.e., $CO_2(GNP_0 + 1)$. The difference $CO_2(GNP_0 + 1) - CO_2(GNP_0)$ is a "slope approximation." (Recall that for a linear function $y = f(x)$, if x increases one unit then the slope tells you how much y changes; also recall that since GNP is expressed in billions of dollars, increasing GNP by one billion means $GNP + 1$). Now write a verbal interpretation of the current rate of increase of CO_2 with respect to GNP.

18.2 GNP AND CO_2 CONTRIBUTORS

Car Groups

Do you think that the GNP affects the gasoline consumed in the United States? Probably, since we drive around to malls to spend those dollars, drive to work, and to the beach if it's still there. In this study we examine the relationship between GNP and gasoline consumption from automobiles.

Recall the function $M(t)$ from Section 8.1; to complete this study, follow the steps in the box below.

1. Adjust the $GNP(t)$ function to get a new $GNP(t)$ function where $t = 0$ in 1940.
2. Find the inverse of the adjusted $GNP(t)$ function.
3. Write the composite $M(GNP)$, which defines number of gallons of gasoline burned in terms of GNP (use the second part of the piecewise function $M(t)$).
4. Graph the function $M(GNP)$.
5. Use your $GNP(t)$ function to determine the current GNP. Use this to compute the "slope approximation" for $M(GNP)$ this year, and write a verbal interpretation.

GNP is also related to the number of cars on the road. Recall the function $A(t)$ from Section 8.1.

6. Write the composite $A(GNP)$, which defines number of automobiles in the United States in terms of GNP.
7. Compute the "slope approximation" for $A(GNP)$ for the current year, and write a verbal interpretation.

Energy Groups

It should be obvious that GNP and energy consumption are related. A lot of energy is required to keep up with the increasing demand for production of goods. In this study we examine the relationship between GNP and energy consumption. Recall the total energy consumption function $TEC(t)$, $t=0$ in 1970 (Section 9.4). To complete this study, follow the steps below.

1. Adjust the $GNP(t)$ function to get a new $GNP(t)$ function where $t=0$ in 1970.
2. Find the inverse of the adjusted $GNP(t)$ function.
3. Write the composite $TEC(GNP)$, which defines total United States energy consumption in terms of GNP.
4. Graph the function $TEC(GNP)$.
5. Use your $GNP(t)$ function to determine the current GNP. Use this to compute the "slope approximation" for $TEC(GNP)$ for this year, and write a verbal interpretation.

Rain Forest Groups

Even though most of the rain forests are not in the United States, there is a strong relationship between what we consume here and the acreage cut in Brazil or Thailand. Your hamburger may come from cattle which graze on cut rain forest land, and your teak table may come from rain forest in Thailand. In this study we examine the relationship between GNP and carbon emission from deforestation. Recall the functions $L(t)$ and $CG(t)$, where $t=0$ in 1940 (Sections 11.1 and 11.2). To complete this study, follow the steps in the box below.

1. Adjust the $GNP(t)$ function to get a new $GNP(t)$ function where $t=0$ in 1940.
2. Find the inverse of the adjusted $GNP(t)$ function.
3. Write the composite $L(GNP)$, which defines the amount of rain forest cut for logging in terms of GNP.
4. Graph the function $L(GNP)$.
5. Use your $GNP(t)$ function to determine the current GNP. Use this to compute

the "slope approximation" for $L(GNP)$ for this year, and write a verbal interpretation.

6. Write the composite $CG(GNP)$, which defines the amount of rain forest cut for cattle grazing in terms of GNP.
7. Graph the function $CG(GNP)$.
8. Compute the "slope approximation" for $CG(GNP)$ for this year, and write a verbal interpretation.

PART V

Alternate Energy and New Trends: Reducing Carbon Dioxide Emission

INTRODUCTION

You have, at this point, completed a fairly thorough study of carbon dioxide emission from three major sources: automobiles in the U.S., power consumption, and deforestation in the tropics. You have also seen the effect of the overall accumulation of CO_2, in particular, the predicted rise in ocean levels and the possible resulting land loss.

In the following chapters, we will study ways to decrease the emission of carbon dioxide in order to slow the resulting temperature increase. Chapter 20 studies the use of alternative energy sources and other ways of using traditional ones in order to reduce CO_2 emission, and Chapter 21 invites students of *Earth Algebra* to formulate their own solutions to the problem of carbon dioxide emission.

New mathematical topics needed for Part V are linear inequalities and linear programming, which are covered in Chapter 19. Any of the mathematical concepts and methods studied in this text can and should be incorporated into your own plan for CO_2 reduction.

CHAPTER 19

Linear Inequalities in Two Variables and Systems of Inequalities

19.1 GRAPHING LINEAR INEQUALITIES IN TWO VARIABLES

A linear inequality in the two variables x and y looks like

$$ax + by \leq c,$$

or

$$ax + by < c,$$

or

$$ax + by \geq c,$$

or

$$ax + by > c,$$

where a, b, and c are constants. A solution to an inequality is any pair of numbers x and y which satisfy the inequality. Here's an example.

Note: Chapter 19 is a prerequisite for Chapters 20 and 21.

EXAMPLE 19.1

Determine the solution set of $5x + 2y \leq 17$.

One solution to this is $x = 2$ and $y = 3$ because

$$5(2) + 2(3) = 16,$$

which is indeed less than or equal to 17. A pair of numbers which does not form a solution is $x = 3$ and $y = 2$, because

$$5(3) + 2(2) = 19,$$

which is not less than or equal to 17. The pair $x = 2$ and $y = 3$ isn't the only solution; as a matter of fact, there are infinitely many solutions.

Since we can't write down all possible solutions to a linear inequality, a good way to describe the set of solutions to any linear inequality is by a graph. If the pair of numbers x and y is a solution, then think of this pair as a point in the plane, so the set of all solutions can be thought of as a region in the xy-plane. We return to Example 19.1 in order to illustrate how to determine this region. First, solve the inequality for y in terms of x.

$$5x + 2y \leq 17$$

$$2y \leq -5x + 17$$

$$y \leq -\frac{5x}{2} + \frac{17}{2}.$$

Next, graph the line

$$y = -\frac{5x}{2} + \frac{17}{2}.$$

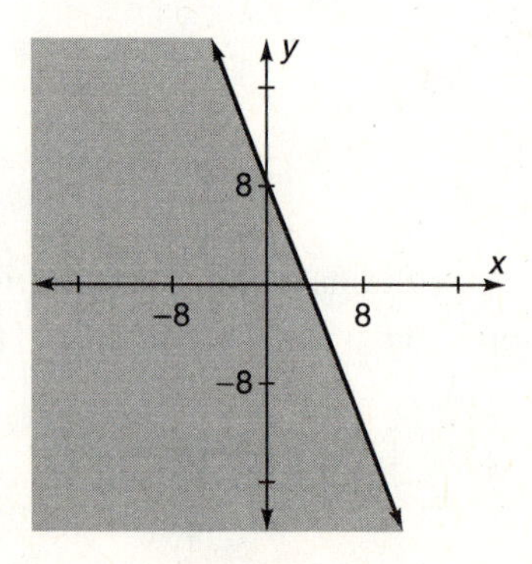

Figure 19.1

The set of points (x, y) which lie on this line is the set of all (x, y) such that y is exactly equal to $-\frac{5x}{2}+\frac{17}{2}$. These points make up part of the set of solutions to the inequality, but not all. We see that y can be also be less than $-\frac{5x}{2}+\frac{17}{2}$, so all points below the line would also be solutions. The shaded region in Figure 19.1 shows the solution set.

EXAMPLE 19.2

Graph the solution set for the inequality

$$3x - 8y \geq 12.$$

Solve for y:

$$-8y \geq -3x + 12$$

$$y \leq \frac{3x}{8} - \frac{3}{2}.$$

Do you notice that the inequality is reversed? That's because we divided by a negative number—any time you multiply or divide an inequality by a negative number, you must reverse the inequality.

Next, graph the line

$$y = \frac{3}{8}x - \frac{3}{2}.$$

Points on this line are part of the solution set, the other part consists of all points below the line. See the shaded region in Figure 19.2.

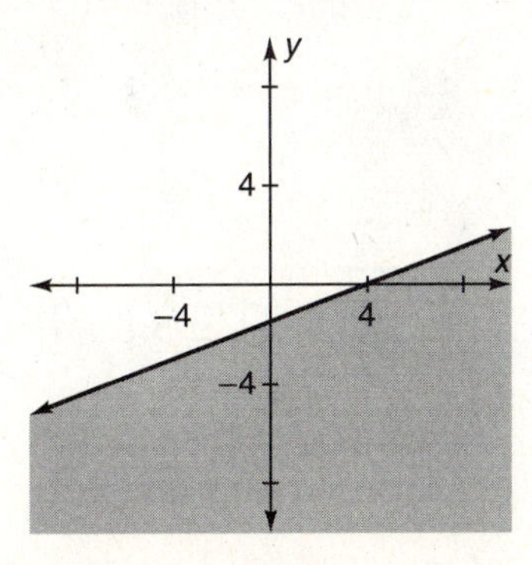

Figure 19.2

EXAMPLE 19.3

Graph the solution set for the inequality

$$-10x - 2y > 7.$$

Solving for y gives

$$-2y > 10x + 7$$

$$y < -5x - \frac{7}{2}. \text{ (Remember the minus?)}$$

This example is a little different because there's no "equals mark" in the inequality. But you still graph the line

$$y = -5x - \frac{7}{2},$$

except draw it as a dotted line. This indicates that the line itself is not part of the solution set. The actual solution set consists of all points below the dotted line. This is because y must be strictly less than $-5x - \frac{7}{2}$. See Figure 19.3.

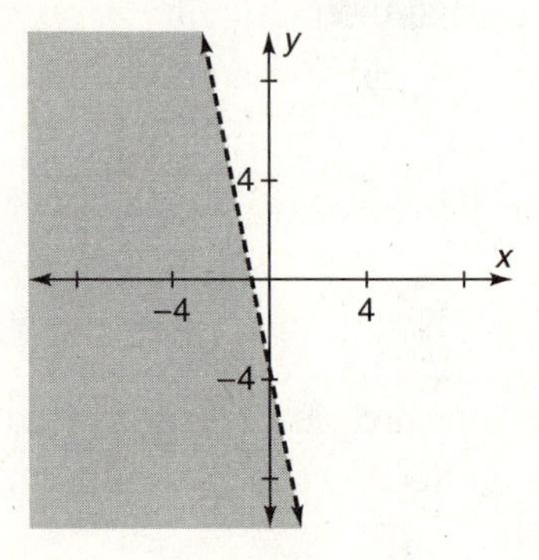

Figure 19.3

19.2 SYSTEMS OF LINEAR INEQUALITIES IN TWO VARIABLES

A system of linear inequalities is a set of one or more linear inequalities; a solution is a pair of numbers (x, y) which satisfies all of the inequalities.

EXAMPLE 19.4

Determine the solution set to the system of linear inequalities

$$x + 5y \leq 20$$

$$3x + 2y \leq 21.$$

The pair of numbers $x = 1$, $y = 2$ is one solution because

$$1 + 5(2) = 11 \leq 20$$

$$3(1) + 2(2) = 7 \leq 21.$$

The pair $x = 0$, $y = 5$ is not a solution because it doesn't even satisfy the first inequality:

$$0 + 5(5) = 25,$$

which is not less than or equal to 20. Notice that it does satisfy the second inequality, but in order to be a solution, it must satisfy both.

As before, a system can have an infinite number of solutions, so we present its solution set by a region in the plane. To illustrate this, we continue with Example 19.4,

$$x + 5y \leq 20$$

$$3x + 2y \leq 21.$$

Solve each inequality for y:

$$y \leq -\frac{1}{5}x + 4$$

$$y \leq -\frac{3}{2}x + \frac{21}{2}.$$

Graph each of the lines on the same coordinate system. The solution set for the first inequality lies in the region below and on the line $y = -\frac{1}{5}x + 4$ and the solution set for the second inequality lies in the region on and below the line $y = -\frac{3}{2}x + \frac{21}{2}$. The solution set for the system lies in the region common to both, and is the darker region shown in Figure 19.4.

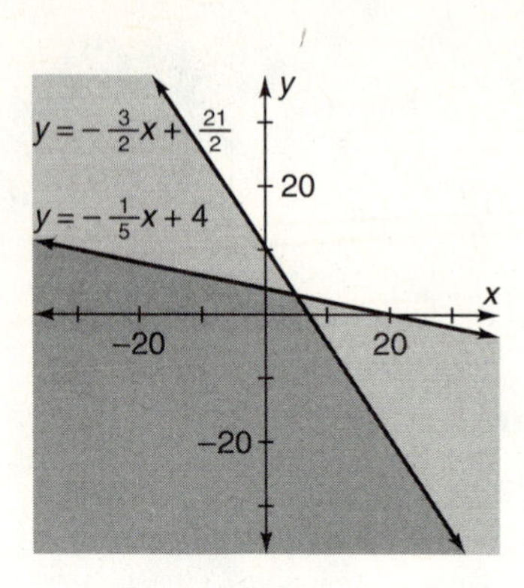

Figure 19.4

The "corner" of this region is the intersection of the two lines, and "corners" will be very important in the next chapter. In order to find this point, return to the original system of inequalities and replace the "inequalities" with "equal marks". The resulting two equations are the equations of the two lines

$$x + 5y = 20$$

$$3x + 2y = 21.$$

This is a system of two linear equations in two unknowns, and its solution is precisely the point of intersection of the two lines. You know how to solve this system: set up the coefficient matrix

$$A = \begin{bmatrix} 1 & 5 \\ 3 & 2 \end{bmatrix},$$

and constant matrix

$$B = \begin{bmatrix} 20 \\ 21 \end{bmatrix},$$

and solve on your calculator. The solution matrix is

$$A^{-1}B = \begin{bmatrix} 5 \\ 3 \end{bmatrix},$$

so $x = 5$ and $y = 3$. The point of intersection, or corner of the region, is (5, 3).

EXAMPLE 19.6

Graph the solution set to the system

$$x + 4y \leq 20$$

$$3x + 4y \geq 28$$

$$x - y \leq 7$$

and find all corners.

As before, graph each line and shade the appropriate region; the solution set to the entire system is the region common to all three, and is shown in Figure 19.7, complete with corners.

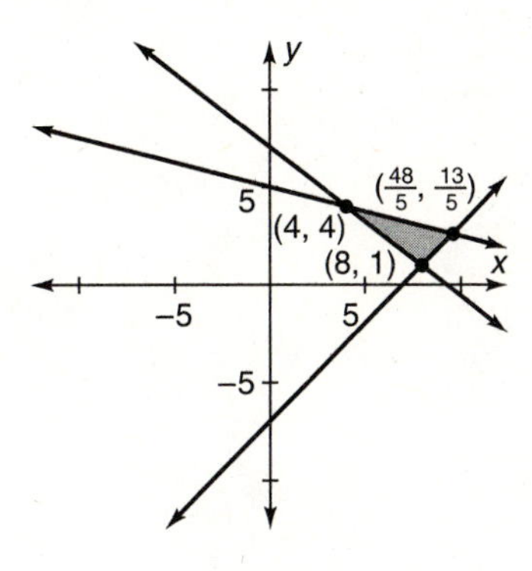

Figure 19.7

19.3 THINGS TO DO

For 1–5, graph the given inequality.

1. $5x + 2y \geq 12$.
2. $2x - 8y < 6$.
3. $0.1x - 2.5y \geq 1.4$.
4. $12x - 11y \leq 65$.
5. $10.2x + 4.5y \leq 14.1$.

For 6–10, graph the system of linear inequalities, and find all corners.

6. $x + y \geq 10$

   $2x - y > 2.$

7. $100x + 400y \leq 750$

   $75x + 250y \leq 700$

   $x \geq 0$

   $y \geq 0.$

8. $x - y \leq 10$

   $2x + y \geq 12.$

9. $1.5x + 2.4y \leq 3.9$

   $3.2x + 10.2y \leq 14.1$

   $x \geq 0$

   $y \geq 0.$

10. $x - y \geq 0$

    $x - 2y \leq 2$

    $x + y \leq 2.$

CHAPTER 20

Cost and Efficiency of Alternate Energy Sources

The techniques used in the studies in this chapter are known to mathematicians as *linear programming* methods. Generally, a linear programming problem involves a particular function, whose maximum (or minimum) value is of interest. But other restrictions apply to the situation; these appear as inequalities. The function to be maximized (or minimized) is called the *objective function,* and the inequalities are called the *constraints*. The studies below seek to maximize utilization subject to cost restriction and imposed carbon dioxide emission levels. This may seem vague, so we work through one particular study.

20.1 METHANOL VERSUS GASOLINE

In this section, we will look at how gasoline-burning cars and methanol-burning cars affect the environment. Methanol emits less CO_2, costs less per gallon, but is less fuel efficient (that is, you get fewer miles per gallon). Suppose some cars burn gasoline and others burn methanol. We will show you how to answer the following question: what combination will let the total population drive the most number of miles without exceeding a fixed cost or emitting more than a fixed amount of carbon dioxide?

First you need some information:

1. when burned, gasoline emits 20 lbs. of CO_2 per gallon,
2. when burned, methanol emits 9.5 lbs. of CO_2 per gallon,
3. it takes 75% more methanol than gasoline to go one mile,
4. the average cost of gasoline is $1.00 per gallon, and
5. the average cost of methanol is $.60 per gallon.

(Numbers 4 and 5 are assumptions based on information current at the time of this writing.)

Note this: the fuel efficiency difference is what makes this a non-trivial problem. Otherwise the solution is easy—switch to methanol! A gallon of methanol costs less and emits less CO_2, but it takes more to go each mile.

Remember that we are trying to maximize the total number of miles which can be driven in the United States in a particular year, so we need the function which describes this. There are two choices of fuel which we consider, gasoline and methanol, and the overall problem is to determine the number of miles which should be driven on each in order to obtain this maximum (subject to cost and emission limitations). The variables in this study are

x = number of miles driven on gasoline (in billions),

y = number of miles driven on methanol (in billions).

Hence the function to be maximized is

$$TMD = x + y,$$

where TMD is the total number of miles ($\times 10^9$) which will be driven by all cars in the U.S. in a particular year. This is the *objective function.*

Next, we discuss the limitations of cost and emission. What limitations would be reasonable? In order to decide this, we look slightly into the future (somewhat arbitrarily, we choose 1995) to see what cost and emission are predicted by our derived models (Chapter 8) which are based on gasoline usage only.

The total number of miles driven by all cars in 1995 ($t = 55$) is projected to be:

$$M(55) \times A(55) \times 10^9 = 1872 \times 10^9;$$

the fuel efficiency is predicted to be:

$$MPG(55) = 19.4;$$

and therefore the number of gallons of gasoline consumed is

$$\frac{1872 \times 10^9}{19.4} = 96 \times 10^9.$$

At a cost of \$1 per gallon, total fuel expenditures are $\$96 \times 10^9$. With an emission of 20 lbs. of CO_2 per gallon, total emission is 1920×10^9 lbs. of CO_2.

Our goal is to emit a little less CO_2, say 1850×10^9 lbs., at not much more expenditure, say $\$100 \times 10^9$. This gives two constraints. One constraint is a CO_2 constraint:

$$\text{total } CO_2 \leq 1850 \times 10^9 \text{ lbs.};$$

and the other is a cost constraint:

$$\text{total cost} \leq 100 \times 10^9 \text{ dollars.}$$

Recall your answers to Exercises 5, 7, 9, and 10 at the end of Section 8.2.

5. Gallons of gasoline per mile = 0.052.
7. Pounds of CO_2 per mile (gasoline) = 1.04.
9. Gallons of methanol per mile = 0.091.
10. Pounds of CO_2 per mile (methanol) = 0.865.

If x miles are driven on gasoline, and y miles are driven on methanol, then total CO_2 emission from both is

$$1.04x + 0.865y \text{ (in billions of pounds).}$$

Our imposed limitation on CO_2 emission is 1850×10^9 lbs, so the constraint is expressed by the inequality

$$1.04x + 0.865y \leq 1850.$$

The second constraint is cost. At a cost of \$1.00 per gallon, gasoline costs \$.052 (rounded) per mile. Methanol, at a cost of \$.60 per gallon, will cost (.60) (.091) = \$.055 (rounded) per mile. So if x miles are driven on gasoline, and y miles are driven on methanol, then the total cost will be

$$.052x + .055y,$$

and with the imposed cost limitation of 100×10^9 dollars, this constraint is expressed by the inequality

$$.052x + .055y \leq 100.$$

There are two more constraints: $x \geq 0$ and $y \geq 0$. (You wouldn't drive a negative number of miles, now would you?) The chart below (20.1) summarizes our work.

Chart 20.1

	CO_2 per Mile (lbs)	Cost per Mile	($) Miles ($\times 10^9$)
Gasoline	1.04	.052	x
Methanol	.865	.055	y
Limitations		1850 ($\times 10^9$)	100 ($\times 10^9$)

Here is the problem:

Maximize the function

$$TMD = x + y$$

subject to the constraints

$$1.04x + .865y \leq 1850,$$

$$.052x + .055y \leq 100,$$

$$x \geq 0, y \geq 0.$$

That means that we have to find, among all the values for x and y that satisfy the inequalities, those which make $TMD = x + y$ as large as possible.

Up until now, all our work has been just setting up the problem; finally, we are ready to find its solution. First, graph the solution set to the system of inequalities (constraints) and find the corners. The graph looks like Figure 20.1.

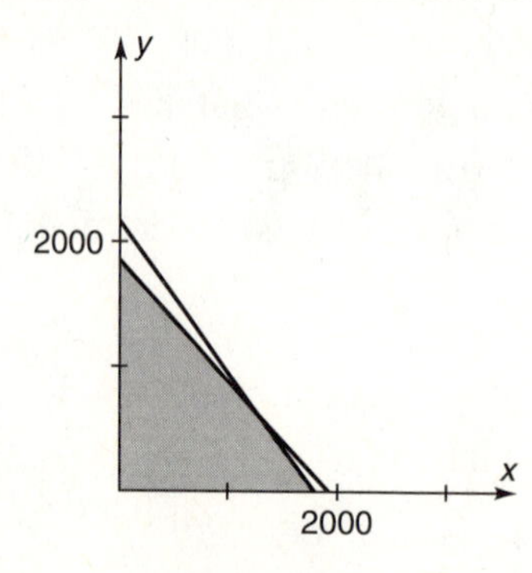

Figure 20.1

The shaded area, called the *feasible region*, shows which points satisfy all the constraints. That's a lot of points, and you have to figure which one makes $x + y$ biggest. Fortunately, something really good happens. **The maximum value of the objective function, if there is a maximum, occurs at one of the corners.** See Figure 20.2.

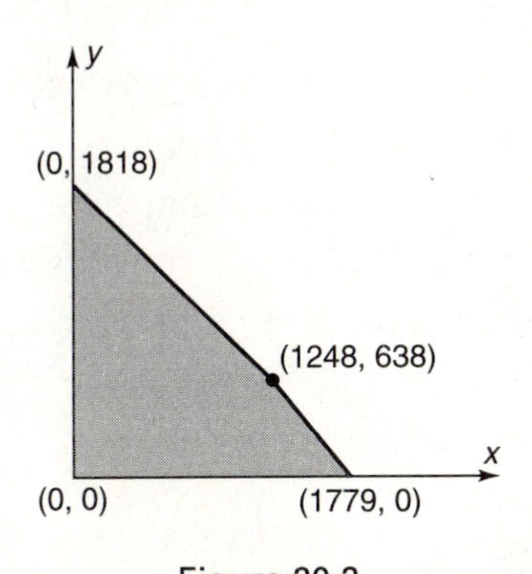

Figure 20.2

Which corner point gives the greatest value for the objective function

$$TMD = x + y?$$

Substitute the values for x and y at each corner to get these answers:

$$0 + 1818 = 1818$$

$$1248 + 638 = 1886$$

$$1779 + 0 = 1779$$

$$0 + 0 = 0$$

The maximum value occurs at the corner (1248, 638). What does that mean? If methanol and gasoline are both available for use in U.S. automobiles, and if 1248×10^9 miles are traveled on gasoline and 638×10^9 miles on methanol, the U.S. drivers can get the maximum number of total miles, which is 1886×10^9 miles. This can be done without exceeding the limit of 1850×10^9 lbs. of CO_2 emission and 100×10^9 dollars for fuel.

That was complicated, so we'll summarize the steps:

1. clearly identify your variables;

2. determine the objective function;

3. determine the constraints (these are the inequalities);

4. graph the system of inequalities to determine the shaded region which is the feasible region;
5. find the "corners," or vertices, of the feasible region;
6. evaluate the objective function at each of these corners. The maximum or minimum will occur at a corner, if there is a maximum or minimum.

Finally, a note on why you only need to look at the corners when you are finding the maximum or minimum value of the objective function. Of course, $x + y$ can be as large as you want, but the problem is to stay inside the feasible region. The picture (Figure 20.3) shows $x + y = 3000$, $x + y = 2000$, and the one that passes through the corner at which the solution occurs. The latter line is $x + y = 1886$. None of the points on the line $x + y = 3000$ are inside the feasible region, so none of these points are valid solutions. $x + y = 2000$ is closer. Notice that all the lines

$$x + y = \text{any number}$$

are parallel. Slide the highest line down, staying parallel until you first hit a point in the feasible region. What point did you hit? A corner! So where is the maximum? At a corner! (See Figure 20.3.)

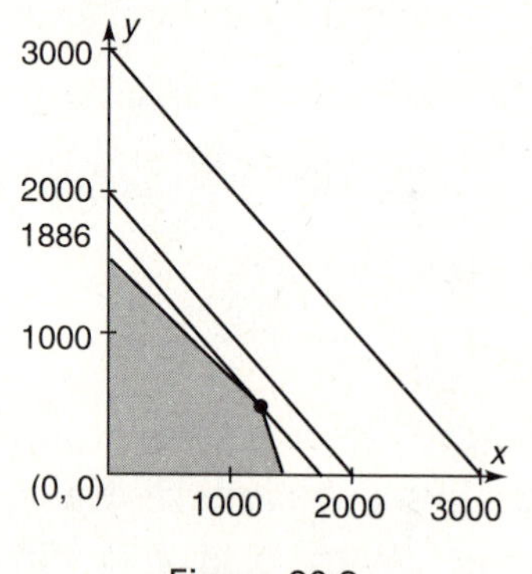

Figure 20.3

Now here's one for you that's almost the same. Using the same information as above, maximize the miles which can be driven if total costs are allowed to rise to $\$110 \times 10^9$ and CO_2 emissions can't exceed 1800×10^9 pounds.

Now try this: you want to improve the rate of CO_2 emission by 10% of the projected level for 1995, so you set a limit of 1750×10^9 lbs. of CO_2. To do this, you are willing to pay a little more for fuel, and you set a limit of $\$105 \times 10^9$. To further encourage the use of alternative fuels, you (having great power) raise the price of gasoline to \$1.10 per gallon while keeping the price of methanol at \$.60 per gallon. What is the maximum number of miles which could be driven?

20.2 WHAT ARE SOME OTHER ALTERNATIVES?

Now that you have seen how linear programming works, here is another study which can be resolved using these methods. This is to be a group effort.

You can use the techniques of linear programming to determine how to reduce the amount of carbon emitted from the consumption of power by changing from coal to natural gas. Almost half of the energy consumed in the United States is from these two sources. Annual coal consumption is about 18.8×10^9 million BTU, while natural gas consumption is 18.6×10^9 million BTU. Coal emits more carbon than natural gas, but it's cheaper. For every million BTU of coal consumed, 25 kilograms of carbon are emitted, while the equivalent consumption of natural gas emits only 14.5 kilograms of carbon. If there were no economic considerations, and an unlimited supply of natural gas, it would make sense to switch entirely from coal to natural gas. But there are economic considerations, and one is the relative prices. A million BTU of coal costs \$1.413, while a million BTU of natural gas costs \$1.538. Here's what we're going to do: first, we'll see what the current situation is, with regard to power consumption, carbon emission, and cost.

In order to determine reasonable limitations to impose on costs and carbon emission, we look at the situation in 1991, then determine how much power we can use if we restrict carbon emission but are willing to pay a little more. This can be determined by completing Chart 20.2A on page 210.

After completing chart 20.2A you see that total cost is $\$55.1712 \times 10^9$ and total carbon emission is $739.7\text{kg} \times 10^9$.

Let's decrease the carbon emission by just over 10%, to 650×10^9 kg, and accept an increase in cost of just under 10%, say to $\$60 \times 10^9$. We're ready to state the problem. Chart 20.2 B on page 210 should help you.

Chart 20.2A

Fuel	Total Consumption (10^9 million BTU)	1991 Consumption		Carbon Emission per million BTU	Total Carbon (kg x 10^9)
		Cost per million BTU	Total cost ($\$ \times 10^9$)		
Coal	18 .8	\$1.413	?	25 kg	?
Natural Gas	18.6	\$1.538	?	14.5 kg	?
		Total cost:	?	Total Emission:	?

What is the maximum amount of energy from coal and natural gas that can be consumed subject to the constraints that the cost is no more than $\$60 \times 10^9$ and the carbon emitted is at most 650×10^9 kg? Assume that you can't change either the prices or the amount of carbon emitted per million BTU. Do this just like we did the methanol/gasoline problem.

Chart 20.2B

Fuel	Cost (\$ per million BTU)	Carbon (kg per million BTU)	Consumption (million BTU $\times 10^9$)
Coal	1.413	25.0	x
Natural Gas	1.538	14.5	y
Limitations	60 ($\times 10^9$)	650 ($\times 10^9$)	

What would happen if the price of coal goes up to \$1.50 per million BTU and the total cost is allowed to reach $\$62 \times 10^9$?

CHAPTER 21

Creating New Models for the Future

21.1 A NEW MODEL FOR FUTURE CARBON DIOXIDE EMISSION

In this chapter, you have the opportunity to formulate your own plan to decrease projected carbon dioxide emission. We show you one example of how this might work, and then suggest other similar ways to accomplish this. The example below was done in 1991 and reductions in projected CO_2 emission were made for the following ten year period.

What if you had the power to slow the increase of atmospheric carbon dioxide concentration so that the actual emission is 10% less in 2001 than the predicted amount for that year? That's a ten-year period from 1991, so there is time to accomplish this relatively smoothly. You can't just command everyone to stop driving, or maybe just drive halfway to work, or to stop using electric lights immediately; that obviously will not work. So however you slow emission, it will have to be a gradual process over this ten year period. One nice and simple way to accomplish this goal is to initiate a simple linear reduction of CO_2 emission from 1991 to the year 2001. Here's what we mean by that. Pretend you are very, very powerful and can somehow control CO_2 emission from energy consumption in the U.S., but you are not powerful enough to do anything about rainforest destruction, or anything else that produces carbon dioxide. So, what you can do is gradually, and uniformly, change CO_2 emission from energy consumption in the U.S. so that by the year 2001 a 10% reduction of the predicted overall emission will have been achieved. This may still be an increase over the 1991 emission, but at least it will be less than the predicted amount for 2001.

You need two things to figure out how this reduction will work. The first is the equation for worldwide carbon emission (Chapter 14, Section 14.2).

$$C(t) = 7(1.01)^t,$$

where $C(t)$ = gigatons of carbon emitted in year $1991 + t$, $t \geq 0$. The other is an equation which can be used to predict carbon emission in the U.S. from energy consumption (sources are coal, petroleum, natural gas, etc.):

$$EC(t) = 0.003t^2 + .083t + 2.252,$$

where $EC(t)$ = gigatons of carbon emitted in $1991 + t$, $t \geq 0$. (This model has been derived by the authors, and is provided for free.)

Next, to make up the difference between $C(t)$ and $EC(t)$, introduce a new variable, $OC(t)$, defined to be gigatons of carbon emitted from all sources other than energy consumption, where t is the number of years after 1991, but $t \geq 0$.

This is nice because we get this simple equation

$$* \; EC(t) + OC(t) = C(t).$$

You want to reduce $C(10)$ by 10% by changing the $EC(t)$ term in the equation.

The predicted carbon emission in 2001 is $C(10) = 7.732$ gigatons. Reducing this figure by 10% is the same as multiplying it by .90 (why?), so we get

$$.90(7.732) = 6.959 \text{ gigatons of carbon.}$$

This is your target worldwide emission for 2001. You need to know $OC(10)$ also:

$$\begin{aligned} OC(10) &= C(10) - \mathrm{EC}(10) \\ &= 7.732 - 3.382 \\ &= 4.35 \text{ gigatons.} \end{aligned}$$

This figure must remain constant; you cannot change this.

In equation * we need to replace the $EC(t)$ term by the unknown target emission from energy consumption in the United States in 2001. Denote this unknown by y. To find this target emission y you must solve the equation

$$y + 4.35 = 6.959$$

to get

$$y = 2.609 \text{ gigatons.}$$

It is interesting to note that this is a reduction of approximately 23% in the predicted emissions from energy for 2001.

A straight linear reduction from present EC, i.e., $EC(0)$, means this: first $EC(0) = 2.252$, which corresponds to the point (0, 2.252). Next, $y = 2.609$, which corresponds to (10, 2.609). Write the equation of the line which goes through these two points.

$$\text{Slope } m = \frac{2.609 - 2.252}{10 - 0} = 0.0357.$$

Substitute into the point-slope equation for a line:

$$TE - 2.252 = 0.0357(t - 0)$$

$$TE(t) = 0.0357t + 2.252,$$

where $TE(t)$ = target carbon emission from energy consumption in gigatons for year $1991 + t$, $0 \leq t \leq 10$.

Note that the slope of this linear function is .0357, which is positive, so carbon emission from energy is still growing, but at a smaller rate than previously predicted. You may say that .0357 is a really small number, and so emission is not increasing much. But, remember, we're talking about gigatons of carbon, and a gigaton is BIG!

This last restriction, $0 \leq t \leq 10$, is made because your new equation $TE(t)$, is only applicable during the ten years 1991–2001.

You can now use this reduction equation to project target emission for each year during the relevant time period. Here are the figures for alternate years (Chart 21.1).

Chart 21.1 Target Carbon Emission from Energy Consumption in U.S.

Year	Gigatons of Carbon
1992	2.288
1994	2.359
1996	2.431
1998	2.502
2000	2.573

Shown in Figure 21.1 is the graph of the original emission function together with the new reduction function.

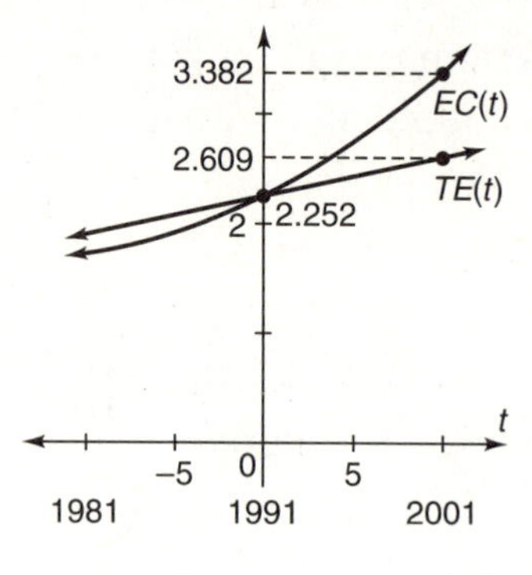

Figure 21.1

21.2 BE PART OF THE SOLUTION, NOT THE PROBLEM

Each group has been assigned (or chosen) one of the sources: automobiles, power, or deforestation. You also have your models which describe total CO_2 or carbon emission from each. Each model involves three factors: "automobiles" has number of cars, miles driven, and MPG; "power" has coal, natural gas, and petroleum; "deforestation" has logging, cattle, and agriculture.

Your group should present a ten year plan to reduce by 23% the predicted carbon or carbon dioxide emission from your source. One way to do this is with a straight linear change of one of the factors involved as shown in Section 21.1 so that in the ten years, CO_2 or carbon from your source is 23% less than its predicted value. You can change only one factor, the others will remain the same. For example, if your source is "automobiles," then you may choose to place requirements on the factor MPG; i.e., require that MPG be increased uniformly over the next ten years so as to achieve the 23% reduction in emission in automobiles. You only change your MPG equation; the "number of cars" and "miles driven" equations will not change.

There are, of course, other ways to reduce carbon emission. Your group may decide to involve more than one factor or to use other types of equations. Be creative.

We suggest that your group prepare both a written and an oral presentation of your ten year plan. Both should include, at least, information such as that provided in the example we presented in Section 21.1.

The following items should be considered in devising your plan and preparing your report: feasibility, originality, clarity of explanation, and mathematical content.

There are many other interesting things you can do with your new reduction model. Be creative. Think of your own.

SUPPLEMENT

Graphing Calculator Manual

INTRODUCTION

The material included in this supplement has been reprinted (with permission) from various calculator manufacturers' manuals for the Casio 7700G, Sharp 9200–9300, TI–81, and TI–85. It includes instruction on most of the calculator operations necessary to do *Earth Algebra*. More detailed information in the use of the calculator, and information on such topics as statistical techniques and programming methods are not included here. For instruction on such topics, or for more details, consult your manual.

We wish to point out that any graphing calculator with matrix capabilities can be used for this course.

Casio fx–7700G

Welcome to the world of Graphing Calculators and the Casio fx-7700G.

Quick-Start is not a complete tutorial, but it will take you through many of the most common functions, from turning the power on through graphing complex equations. When you're done, you'll have mastered the basic operation of the fx-7700G and will be ready to proceed with the rest of this manual to learn the entire spectrum of functions the fx-7700G can perform.

Each step of every example is shown graphically to help you follow along quickly and easily. For example, when you need to enter the number 57, we've indicated it as follows:

Press [5] [7]

Whenever necessary, we've included samples of what your screen should look like. If you find that your screen doesn't match the sample, or in fact you need to start over for any reason, you can do so by pressing the "All Clear" button. [AC]

POWER ON/OFF

To turn your unit on, press [AC] ON

To turn your unit off, press [SHIFT] [AC] OFF

NOTE: *Your unit will automatically shut itself off after six minutes of inactivity.*

ADJUSTING THE CONTRAST

1. Press [MODE]
   The following screen will appear:

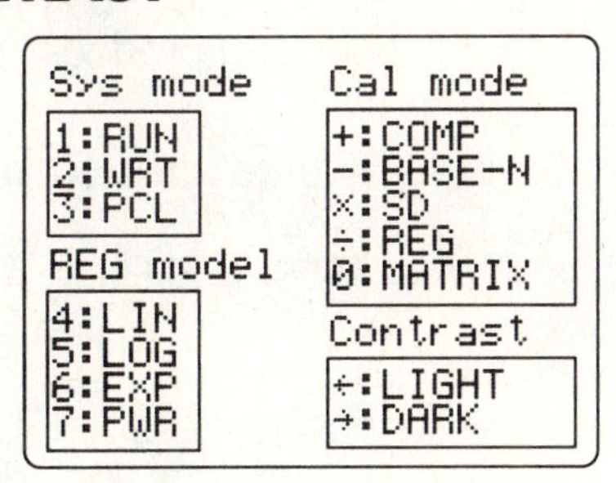

2. Press ◀ to lighten screen or ▶ to darken screen.
3. Press [AC] to clear the screen.

This information is from the *Power Graphic fx–7700G Owner's Manual* published by Casio and is used with permission.

The fx-7700G features a variety of modes that enable you to perform specific functions. To begin this Quick-Start guide, you will need to set the correct system mode and calculation mode.

Setting the system mode

1. After turning the fx-7700G on, press MODE

The following screen will appear:

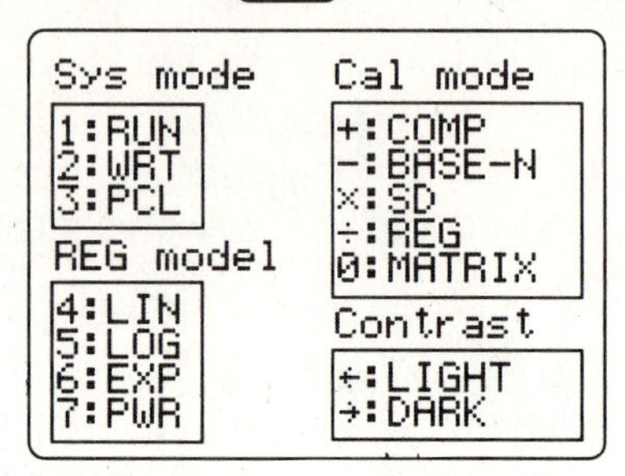

2. Press 1 which corresponds to RUN in the box labelled Sys mode.

The following screen or similar will appear:

You are now in the RUN mode, where you can perform manual computations and produce graphs.

Setting the calculation mode

1. Press MODE 2. Press + which corresponds to COMP in the box labelled Cal mode.

You are now in the COMPUTATION mode, where you can perform general computations, including functional computation.

BASIC COMPUTATIONS

Unlike a regular calculator, which lets you see only one step of your problem at a time, the fx-7700G displays the entire problem on its large, computer-like screen. You enter problems just as you would write them, as you will see in the following example:

EXAMPLE: 15 x 3 + 61

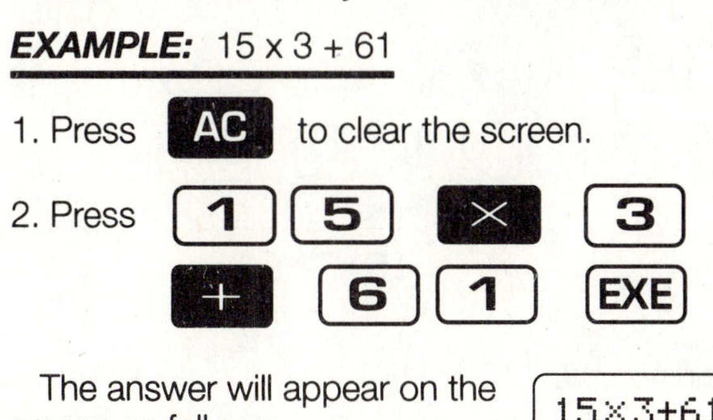

1. Press AC to clear the screen.
2. Press 1 5 × 3 + 6 1 EXE

The answer will appear on the screen as follows:

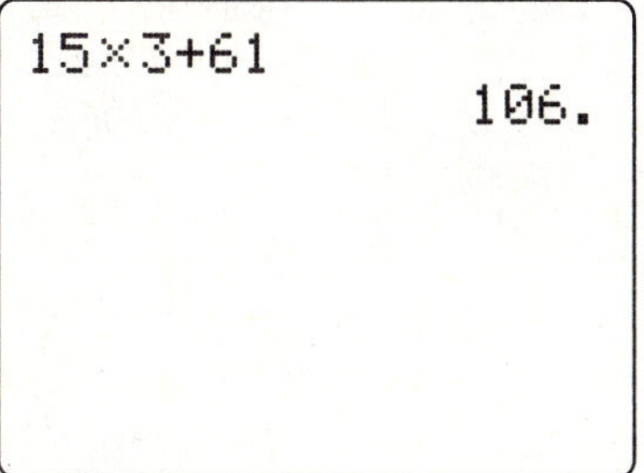

NOTE: *In mixed arithmetic operations, the fx-7700G automatically gives priority to multiplication and division, and computes those operations before addition and subtraction.*

Keep this problem displayed on your screen while you move on to the next example.

Grouping within an equation

You can also group certain operations within your equation using the parentheses keys. [( ] [ )]

EXAMPLE: 15 x (3 + 61)

1. Press [1] [5] [×] [(] [3] [+] [6] [1] [)] [EXE]

The following screen will appear:

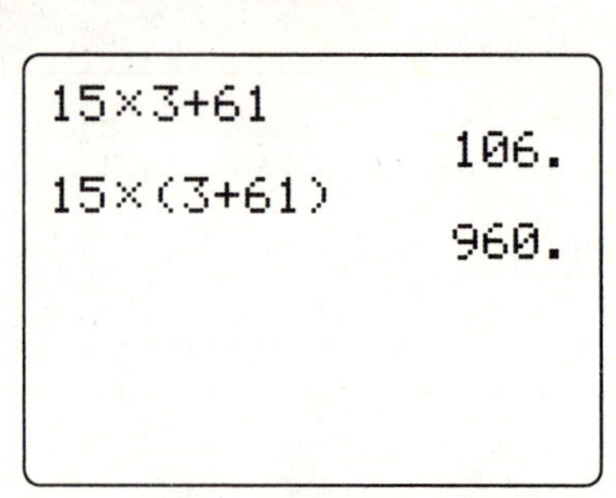

Note that your previous calculation remains on the screen. The new calculation is displayed beneath it for easy comparison.

Now let's try a variation on that problem by positioning the parentheses differently.

EXAMPLE: (15 x 3) + 61

1. Press [(] [1] [5] [×] [3] [)] [+] [6] [1] [EXE]

The following screen will appear:

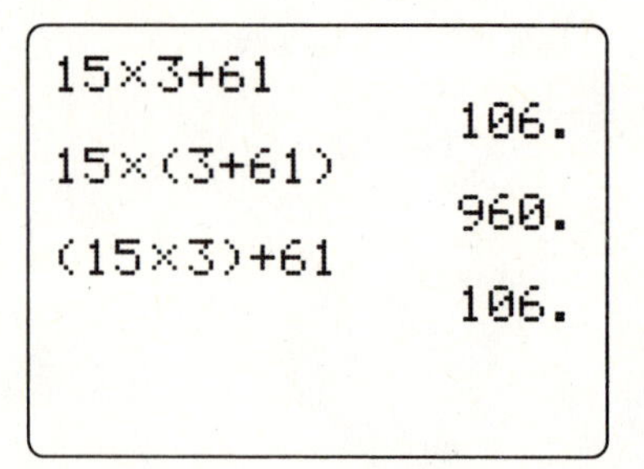

As you can see, the fx-7700G displays all three problems simultaneously.

USING BUILT-IN VALUES

The fx-7700G features several convenient built-in functions and values that you can enter into your equations quickly and easily.

EXAMPLE: 25 x sine of 45 (In Deg mode)

1. Press [AC]
2. Press [2] [5] [×] [sin] [4] [5]
3. Press [EXE] and the answer will appear on the screen as follows:

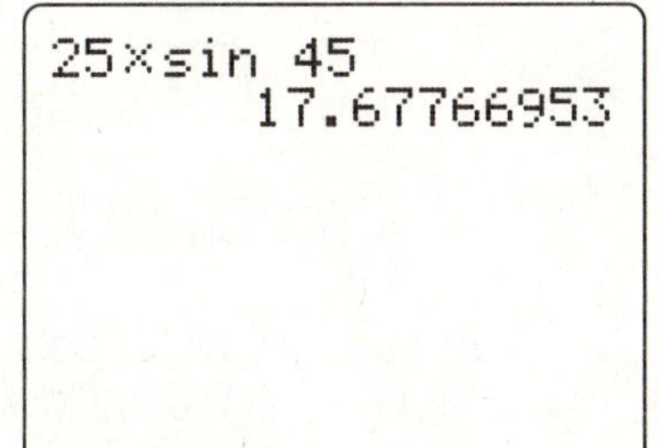

Using the Replay feature

With the replay feature, you can go back in and change any part of your equation at any time, even after the fx-7700G computes the answer, without having to rewrite the entire equation. We'll use the previous equation as an example. Let's say you need to change the sine of 45 to sine of 55, but everything else in the equation remains the same.

1. Press [◀] This will bring you back into the equation.
2. Press [◀] twice so the flashing cursor is on the 4.
3. Press [5] to overwrite a 5.
4. Press [EXE] and the fx-7700G will quickly recompute the new solution:

```
25×sin 55
        20.47880111
```

FRACTIONS

The fx-7700G makes it easy to work with fractions with its fraction key. $a^b/_c$ On screen, the ⌟ symbol is entered between each value of the fraction. For example, $1^{15}/_{16}$ would appear as 1⌟15⌟16

EXAMPLE: $1^{15}/_{16} + {}^{37}/_{9}$

1. Press AC
2. Press 1 $a^b/_c$ 1 5 $a^b/_c$ 1 6 + 3 7 $a^b/_c$ 9 EXE

The answer will appear on the screen as follows:

```
1⌟15⌟16+37⌟9
        6⌟7⌟144.
```

Converting the answer to a decimal equivalent

With the answer still on your screen,

1. Press EXE $a^b/_c$ and the decimal equivalent of your answer (6.048611111) will appear on the screen.

Converting the answer to an improper fraction

With the answer still on your screen,

1. Press EXE SHIFT $a^b/_c$ and your answer (871⌟144) will appear on the screen in the form of an improper fraction.

EXPONENTIALS

Exponentials are another function the fx-7700G can perform quickly and easily.

EXAMPLE: 1250×2.06^5

1. Press AC
2. Press 1 2 5 0 × 2 · 0 6
3. Now you are ready to enter the exponent value. Press the exponent key x^y and x^y will appear on the screen. The number directly preceding the x, in this case 2.06, is the base number.
4. Press 5 The number 5 now appears after the x^y symbol, and represents the exponential value.
5. Press EXE and the answer will appear on the screen as follows:

```
1250×2.06x^y5
      46370.96297
```

GRAPHING

The fx-7700G has the ability to present graphic solutions to a variety of complex equations. But before you can begin you must make sure you are in the correct GRAPH MODE:

Setting the graph mode

1. Press AC MODE SHIFT and the second mode screen will appear:

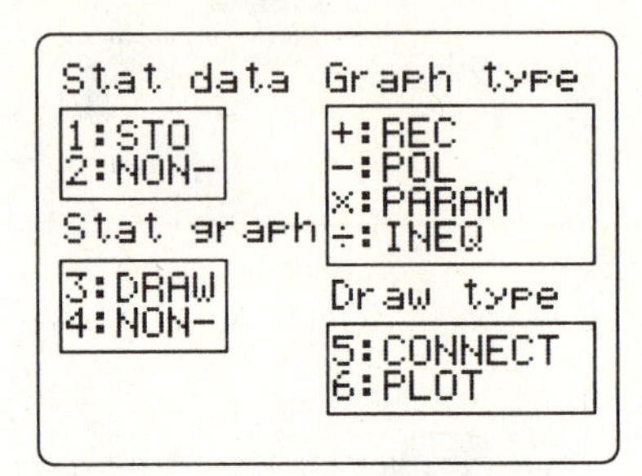

2. Press + which corresponds to REC in the box labelled Graph type, to set the graph mode to rectangular coordinate graph.
3. Press MODE + to set the COMP (computation) mode.

Graphing a built-in function

The fx-7700G can quickly create a graph of one of its built-in values or functions.

EXAMPLE: y = sin x

4. Press Graph
5. Press sin (x is assumed)
6. Press EXE and the following graph will appear:

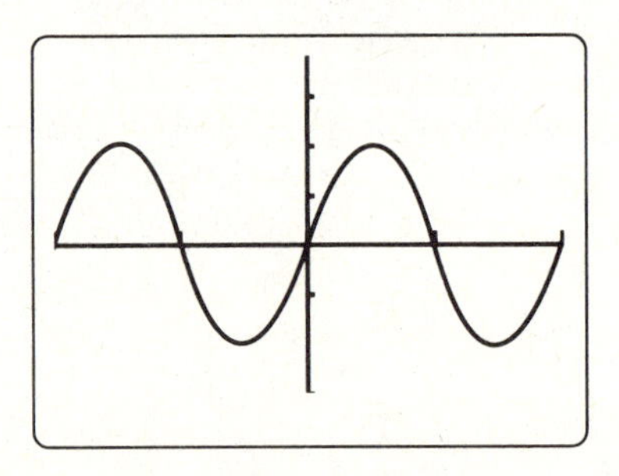

Returning to the equation

If you find that you need to return to your equation to change or replace certain values, you can do so simply by pressing the Graph-Text toggle key. G↔T The fx-7700G has two separate areas of its memory: one for your formula, the other for graphs.

1. Press G↔T once to see the equation, then again to see the graph.

Trace function

The trace function lets you select an exact point on the graph and display the coordinates of that point.

1. With the graph still on your screen, press **Trace** [F1]. The following screen will appear:

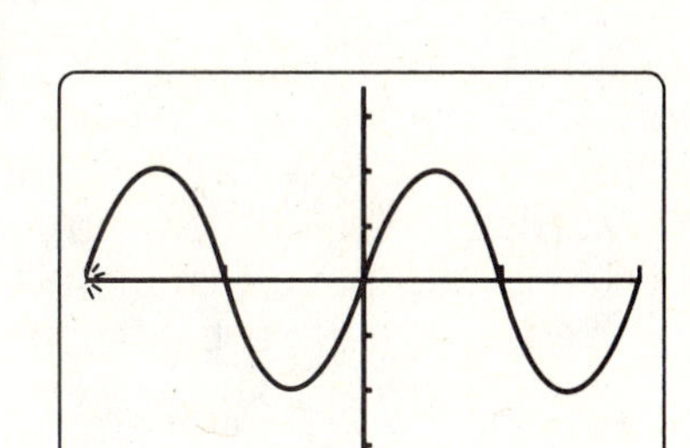

Notice that a cursor has appeared at the left-most point on the X axis and its coordinates have appeared at the bottom of the screen. Move the cursor to the right by pressing the ▶ key, then back to the left using the ◀ key. Pressing the button once will move the cursor one point, while holding it down will cause continuous movement. (The values may be approximated due to the space limitations of the screen.

2. Press **Coord** [F6] to view the full value of the X coordinate in unabbreviated form.
3. Press **Coord** [F6] to view the full value of the Y coordinate in unabbreviated form.
4. Press **Coord** [F6] a third time to see both coordinates simultaneously.
5. Press **Trace** [F1] to exit the trace function.

Scrolling in four directions

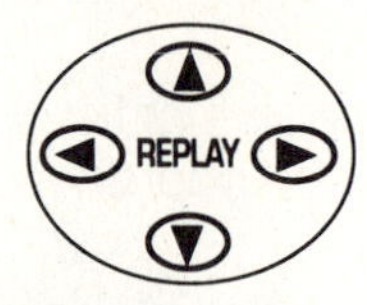

1. Pressing any arrow key lets you scroll to see different sections of your graph.

Returning to your original graph

After scrolling, you needn't retrace your steps to get back to your original graph. You can do it quickly and easily using the function keys (F keys) to enter a selection from one of the many FUNCTION MENUS the fx-7700G employs. A function menu is a group of up to 6 functions that are displayed across the bottom of the screen. To select one of the choices, press the corresponding F key.

2. Using the ▶ key, scroll so the Y axis is at the left of the screen.

3. Press Zoom F2 and the following screen will appear:

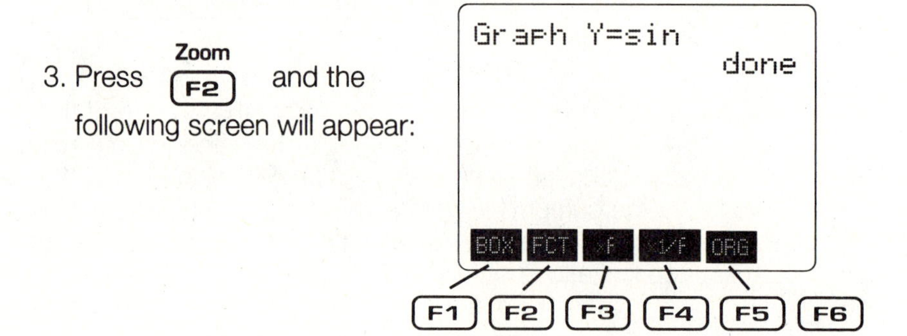

The first five function keys in the function menu each correspond to one of the five boxes along the bottom of the screen. (The sixth function key is inactive in this instance.) The one we'll concern ourselves with now is F5 which corresponds to ORG (original) on the screen.

4. Press F5 to bring you back to your original graph.

Zoom function

Another of the powerful graphing features of the fx-7700G is zooming. This allows you to enlarge a portion of your graph for detailed analysis, or zoom out for a broader view.

Zooming in

1. Press F2
   The following screen will appear:

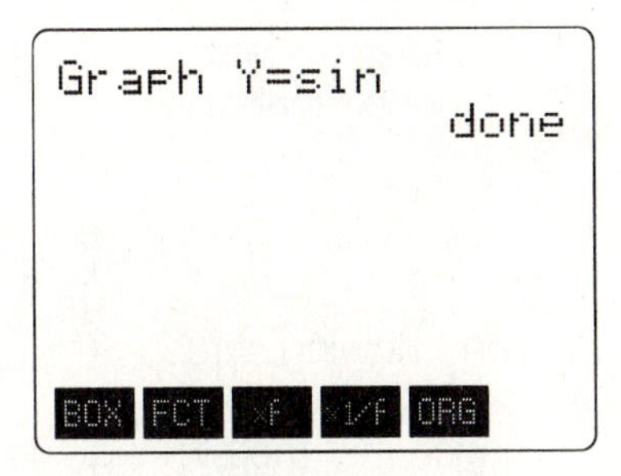

2. Press F3 which corresponds to the ×F box on the screen, to zoom in on your graph. The screen will now show a view that is enlarged by a predetermined factor. (Later in the manual, you'll learn how to set your own factor of enlargement or reduction.)

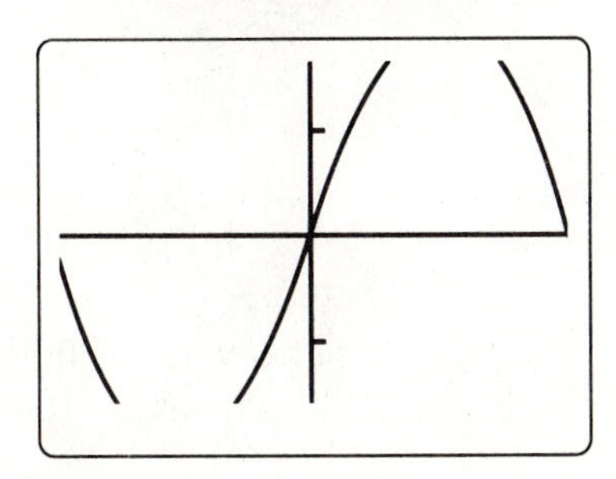

3. Press F2 to show the zoom function menu.

4. Press F5 to return to your original graph.

Zooming out

5. Press F2 to show the zoom function menu.

6. Press F4 which corresponds to × 1/F on the screen, to zoom away from the graph. The screen should now look like this:

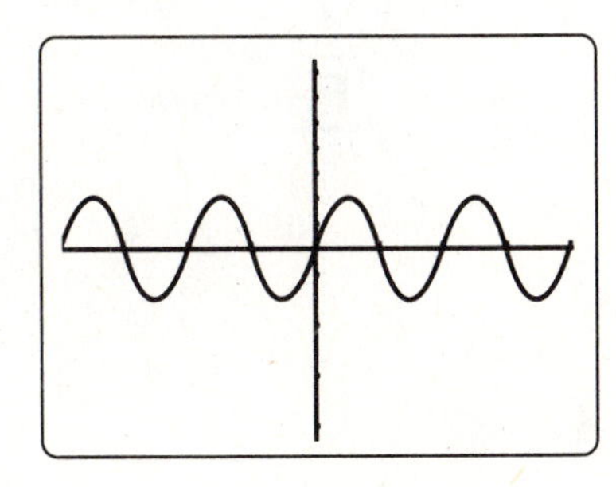

Using the Box function to zoom

This function lets you define any portion of the screen and magnify it for further analysis.

1. Press [F2] to display the zoom function menu.
2. Press [F1] which corresponds to BOX on the screen.

The following screen will appear:
Notice that the blinking cursor is at the origin.

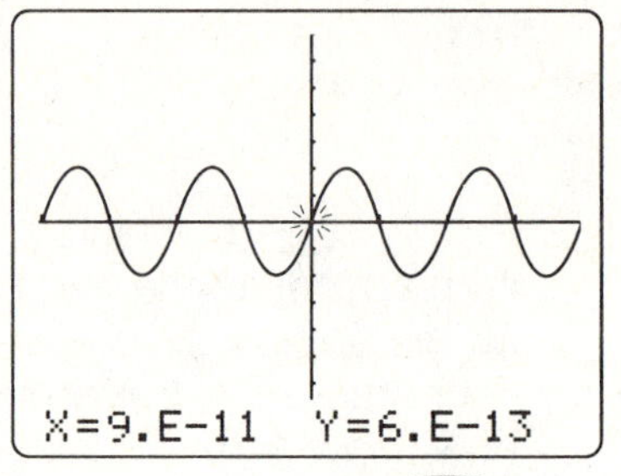

3. Using the arrow keys, move the cursor to a spot which will define one corner of the area, or "box," you wish to zoom in on.

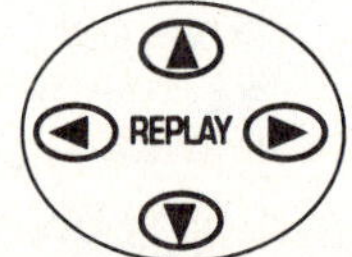

4. Press [EXE] to "anchor" the cursor, creating the first corner of the box. Now, use the arrow keys to draw a box over the area you wish to enlarge.
5. Press [EXE] and the area you defined will enlarge to fill the entire screen.
6. Press [F2] to display the zoom function menu.
7. Press [PRE] twice to clear the zoom function menu.

Creating the graph

EXAMPLE: $y > x^2 - 5x - 5$
$y < x - 2$

3. Press [Graph] and the following screen will appear:

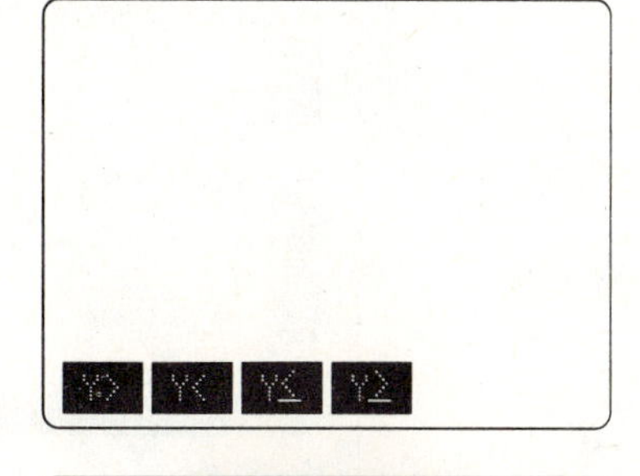

4. Press [F1] which corresponds to the Y> box on the screen.

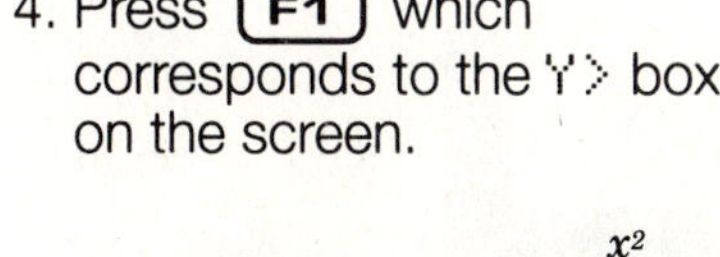

5. Press [X,θ,T] [SHIFT] [x^2] [—] [5] [X,θ,T] [—] [5] [EXE]

The following screen will appear:

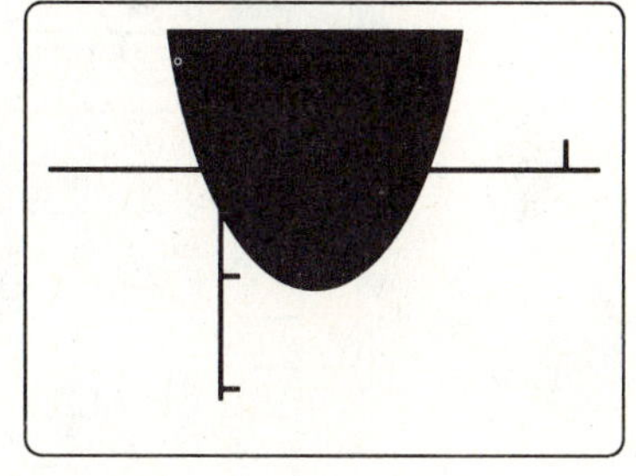

6. Press [Graph] to enter the next inequality.

7. Press [F2] which corresponds to the Y< box on the screen.

8. Press [X,θ,T] [—] [2]

[EXE] The following screen will appear:

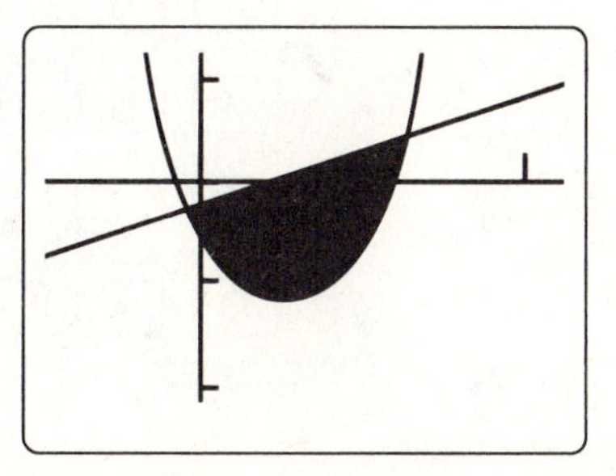

If you've completed this Quick-Start section, you are well on your way to becoming an expert user of the Casio fx-7700G Power Graphic Calculator.

To learn all about the many powerful features of the fx-7700G, read on and explore!

Keys and Their Functions

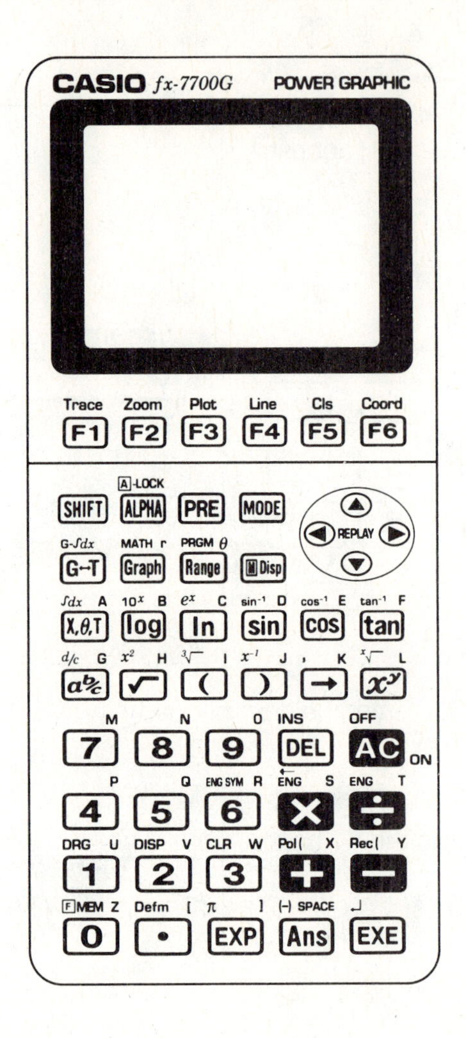

■ The Keyboard

Many of the unit's keys are used to perform more than one function. The functions marked on the keyboard are color coded to help you find the one you need quickly and easily.

Shifted function —— 10^x B —— Alpha function
Primary function —— [log]

• Primary Functions

These are the functions that are normally executed when you press the key.

• Shifted Functions

You can execute these functions by first pressing the [SHIFT] key, followed by the key that is assigned the shifted function you want to execute.

• Alpha Functions

An alpha function is actually the simple input of an alphabetic letter. Press the [ALPHA] key, followed by the key that is assigned the letter you want to input.

Alpha Lock

Normally, once you press [ALPHA] and then a key to input an alphabetic character, the keyboard reverts to its primary functions immediately. If you press [SHIFT] and then [ALPHA], the keyboard locks in alpha input until you press [ALPHA] again.

■ Key Operations

Trace [F1] ~ Coord [F6] Function Keys

• Use these keys to select functions from the menus that appear on the display.

[SHIFT] Shift Key

• Press this key to shift the keyboard and access the functions marked in orange (or green). The [S] indicator on the display indicates that the keyboard is shifted. Pressing [SHIFT] again unshifts the keyboard and clears the [S] indicator from the display.
• This key is also used during display of a Mode Menu to advance to the next Mode Menu screen.

[A]-LOCK [ALPHA] Alpha Key

• Press this key to input a letter marked in pink on the keyboard.
• Press this key following [SHIFT] to lock the keyboard into alphabetic character input. To return to normal input, press [ALPHA] again.

[PRE] Previous Key

• Use this key to backtrack through menus.

Cursor/Replay Keys

- Use these keys to move the cursor on the display.
- After you press the [EXE] key following input of a calculation or value, press [◀] to display the calculation from the end, or [▶] to display it from the beginning. You can then execute the calculation again, or edit the calculation and then execute it.

[MODE] **Mode Key**
- Press this key to display the Mode Menu.

G-∫dx
[G↔T] **Graph-Text Key**
- Press this key to switch between the graph and text screens.
- Press this key following [SHIFT] before entering data for graphing of an integral.

MATH r
[Graph] **Graph Key**
- Press this key before entering a calculation formula for graphing.
- Press this key following [SHIFT] to display the input screen for functions. For full details on this operation, see page 33.

PRGM θ
[Range] **Range Key**
- Use this key to set or check the range of a graph.
- Press this key following [SHIFT] to display the input screen for program commands.
- Press this key following [ALPHA] to enter the letter θ.

[M-Disp] **Mode Display Key**
- Use this key to check the current calculation mode settings. The mode settings remain displayed while this key is depressed.

∫dx A
[X,θ,T] **Variable Key**
- Press this key to input variables X, θ, and T when setting up a graph.
- Press this key following [SHIFT] to input variables for integration calculations.
- Press this key following [ALPHA] to enter the letter A.

10^x B
[log] **Common Logarithm/Antilogarithm Key**
- Press this key and then enter a value to calculate the common logarithm of the value.
- Press [SHIFT][10^x] and then enter a value to make the value an exponent of 10.
- Press this key following [ALPHA] to enter the letter B.

e^x C
[ln] **Natural Logarithm/Exponential Key**
- Press this key and then enter a value to calculate the natural logarithm of the value.
- Press [SHIFT][e^x] and then enter a value to make the value an exponent of e.
- Press this key following [ALPHA] to enter the letter C.

sin⁻¹ D cos⁻¹ E tan⁻¹ F [sin] [cos] [tan] Trigonometric Function Keys

[sin]

- Press this key and then enter a value to calculate the sine of the value.
- Press this key following [ALPHA] to enter the letter D.

[cos]

- Press this key and then enter a value to calculate the cosine of the value.
- Press this key following [ALPHA] to enter the letter E.

[tan]

- Press this key and then enter a value to calculate the tangent of the value.
- Press this key following [ALPHA] to enter the letter F.

[SHIFT][sin⁻¹]

- Perform this operation and then enter a value to calculate the inverse sine of the value.

[SHIFT][cos⁻¹]

- Perform this operation and then enter a value to calculate the inverse cosine of the value.

[SHIFT][tan⁻¹]

- Perform this operation and then enter a value to calculate the inverse tangent of the value.

d/c G [a b/c] Fraction Key

- Use this key when entering fractions and mixed fractions. To enter the fraction 23/45, for example, press 23[a b/c]45. To enter 2-3/4, press 2[a b/c]3[a b/c]4.
- Press [SHIFT][d/c] to display an improper fraction.
- Press this key following [ALPHA] to enter the letter G.

x^2 H [√] Square Root/Square Key

- Press this key to calculate the square root of the next value you enter.
- Enter a value and then press [SHIFT][x^2] to square the entered value.
- Press this key following [ALPHA] to enter the letter H.

∛ I [(] Open Parenthesis/Cube Root Key

- Press this key to enter an open parenthesis in a formula.
- Press [SHIFT][∛] and then enter a value to calculate the cube root of the value.
- Press this key following [ALPHA] to enter the letter I.

x^{-1} J [)] Close Parenthesis/Reciprocal Key

- Press this key to enter a close parenthesis in a formula.
- Press [SHIFT][x^{-1}] and then enter a value to calculate the reciprocal of the value.
- Press this key following [ALPHA] to enter the letter J.

, K [→] Assignment/Comma Key

- Press this key before entering a value memory name to assign the result of a calculation to the value memory.
- Press this key following [SHIFT] to input a comma.
- Press this key following [ALPHA] to enter the letter K.

[x^y] Power/Root Key

- Enter a value for x, press this key, and then enter a value for y to calculate x to the power of y.
- Enter a value for x, press [SHIFT][$\sqrt[x]{\ }$], and then enter a value for y to calculate the xth root of y.
- Press this key following [ALPHA] to enter the letter L.

[0] ~ [9], [•] Numeric Keys and Decimal Key

- Use the numeric keys to enter a value. Enter decimals using the decimal key.
- Following operation of the [ALPHA] key, each of the numeric keys enters the following letters.
  - [ALPHA][7] enters M.
  - [ALPHA][8] enters N.
  - [ALPHA][9] enters O.
  - [ALPHA][4] enters P.
  - [ALPHA][5] enters Q.
  - [ALPHA][6] enters R.
  - [ALPHA][1] enters U.
  - [ALPHA][2] enters V.
  - [ALPHA][3] enters W.
  - [ALPHA][0] enters Z.
  - [ALPHA][•] enters the open bracket [.
- Following operation of the [SHIFT] key, the menus marked in orange (or green) above these keys are accessed.

[SHIFT][FMEM] — **Function Memory Menu**

This key operation displays the menu used for function memory calculations.

[SHIFT][DRG] — **Unit of Angular Measurement Menu**

This key operation displays the menu used for specification of the unit of angular measurement.

[SHIFT][DISP] — **Display Format Menu**

This key operation displays the menu used for specification of the display format for calculation results.

[SHIFT][CLR] — **Clear Menu**

This key operation displays the menu used for clearing memory contents.

[SHIFT][ENG SYM] — **Engineering Symbol Menu**

This key operation displays the menu used for assignment of engineering symbols to values.

[SHIFT][Defm][EXE]

This key sequence displays the status of the program, function, variable, statistic (SD and LR), and matrix memories, along with the remaining number of steps.

For full details on each menu, see the section titled "Basic Set Up", "Basic Operation".

[AC/ON] All Clear/ON/OFF Key

- Press this key to switch power on.
- Press this key while power is on to clear the display.
- Press this key following [SHIFT] to switch power off.

INS
[DEL] Delete/Insert Key

- Press this key to delete the character at the current cursor location.
- Press [SHIFT][INS] to display the insert cursor ([ ]). You can insert characters while the insert cursor is displayed.

Pol(X Rec(Y ENG S ENG T
[+] [−] [×] [÷] Arithmetic Operation Keys

- Input addition, subtraction, multiplication, and division calculations as they are written, from left to right. Press the applicable key to specify an arithmetic operation.
- You can also use the [+] and [−] keys to specify positive and negative values.
- Following operation of the [ALPHA] key, each of these keys enters the following letters.
  - [ALPHA][×] enters S.
  - [ALPHA][÷] enters T.
  - [ALPHA][+] enters X.
  - [ALPHA][−] enters Y.
- Following operation of the [SHIFT] key, the functions marked in orange above these keys are accessed.

[SHIFT][Pol(] — Coordinate Transformation

Use this operation when transforming rectangular coordinates into polar coordinates.

[SHIFT][Rec(] — Coordinate Transformation

Use this operation when transforming polar coordinates into rectangular coordinates.

[SHIFT][ENG] — Engineering Right

Each time you perform this operation, the decimal of the displayed value shifts three decimal places to the right. This results in conversion of the displayed value from one International System unit to another, as shown in the following table.

Power	Prefix	Symbol
10^{18}	exa	E
10^{15}	peta	P
10^{12}	tera	T
10^{9}	giga	G
10^{6}	mega	M
10^{3}	kilo	k
10^{-3}	milli	m
10^{-6}	micro	μ
10^{-9}	nano	n
10^{-12}	pico	p
10^{-15}	femto	f

Example	12.3456 EXE	12.3456	
	1st operation of SHIFT ENG	12.3456E+00	
	2nd operation of SHIFT ENG	12345.6E-03	
	3rd operation of SHIFT ENG	12345600.E-06	
	4th operation of SHIFT ENG	12345600.E-06	(No change)

SHIFT ENG — Engineering Left

Each time you perform this operation, the decimal of the displayed value shifts three decimal places to the left. This results in conversion of the displayed value from one International System unit to another, as shown in the table above.

Example	12.3456 EXE	12.3456	
	1st operation of SHIFT ENG	0.0123456E+03	
	2nd operation of SHIFT ENG	0.000012345E+06	
	3rd operation of SHIFT ENG	0.000000012E+09	
	4th operation of SHIFT ENG	0.000000012E+09	(No change)

π] EXP Exponent/Pi Key

- Use this key when entering a mantissa and exponent. To input 2.56×10^{34}, for example, enter 2.56 EXP 34.
- Press SHIFT π to input the value of π.
- Press this key following ALPHA to enter the closed bracket].

(–) SPACE Ans Answer/(–) Key

- Press this key to recall the most recent calculation result obtained using the EXE key.
- Press SHIFT (–) when entering a negative value.
- Press this key following ALPHA to enter space.

↵ EXE Execute/Newline Key

- Press this key to obtain the result of a calculation. You can press this key following data input, or after a result is obtained to execute the calculation again using the previous result.
- Press SHIFT ↵ to perform a newline operation.

Modes

You can control the operations of the unit by setting certain parameters, which we call *modes*. When you press the AC ON key and switch power on, the display should appear somewhat like the following illustration.

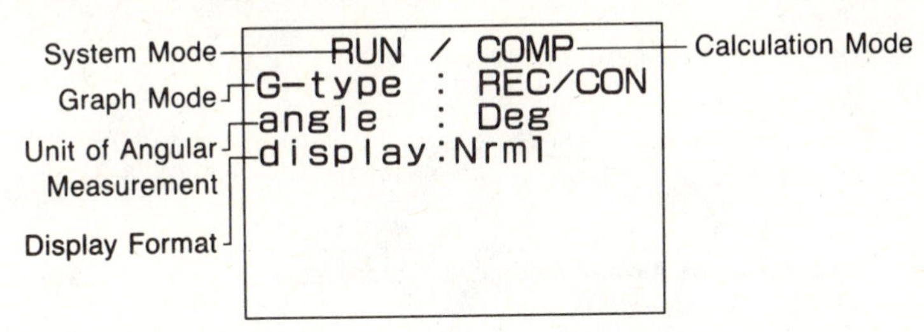

■Using the Mode Menus to Change Modes

There are two mode menus that you can use to change modes. The following explains the content of the menu. The operations that you should perform to change the modes can be found in the applicable sections of this manual.

● To display Mode Menu 1

Press MODE.

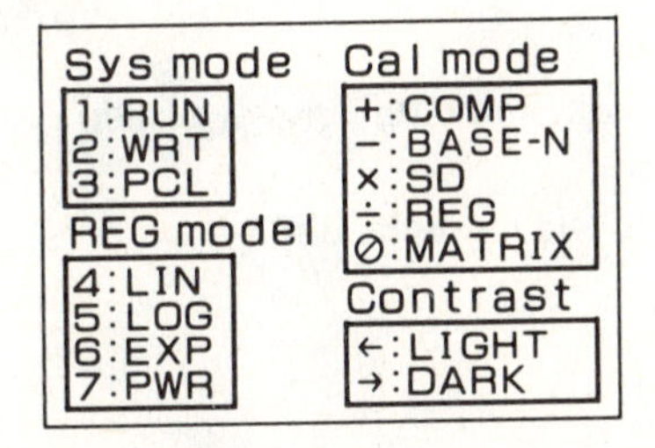

Each of the values and symbols to the left of the mode names stands for key. To select a mode or operation, press the corresponding key.

Sys mode

1: RUN

Use this mode for manual calculations and program execution.

2: WRT

Use this mode for writing or checking programs.

3: PCL

Use this mode to clear programs from memory.

REG model

4: LIN
Use this mode for linear regression.

5: LOG
Use this mode for logarithmic regression.

6: EXP
Use this mode for exponential regression.

7: PWR
Use this mode for power regression.

Cal mode

+: COMP
Use this mode for arithmetic calculations and function calculations. Programs can be executed in this mode.

−: BASE-N
Use this mode for binary, octal, and hexadecimal calculations and conversions.

×: SD
Use this mode for standard deviation calculations.

÷: REG
Use this mode for regression calculations.

0: MATRIX
Use this mode for matrix calculations.

Contrast

←: LIGHT
Press the ◀ key to make the display lighter.

→: DARK
Press the ▶ key to make the display darker.

• To display Mode Menu 2

Press MODE SHIFT.

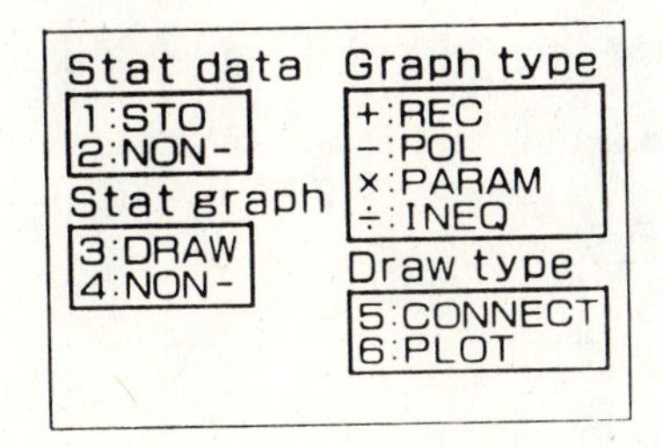

Stat data
1: STO
Use this mode to store statistical data as it is input.

2: NON-
Use this mode if you do not want to store statistical data as it is input.

Stat graph
3: DRAW
Use this mode to draw a statistical graph.

4: NON-
Use this mode if you do not want to draw a statistical graph.

Graph type
+: REC
Use this mode to draw graphs with rectangular coordinates.

−: POL
Use this mode to draw graphs with polar coordinates.

×: PARAM
Use this mode to graph parametrics.

÷: INEQ
Use this mode to graph inequalities.

Draw type
5: CONNECT
Use this mode to connect the points plotted on the graph.

6: PLOT
Use this mode to plot individual (unconnected) points.

• To clear the Mode Displays

Press MODE again.

Basic Set Up

■To specify the Unit of Angular Measurement

Example **To set the unit of angular measurement as degrees**

[SHIFT][DRG] — Display: Deg Rad Gra o r g ([F1])

[F1](Deg)[EXE] — Display: Deg 0.

The following are the units of angular measurement that are available with the unit.

DEG (degrees)	360°	90°
RAD (radians)	2π	$\pi/2$
GRA (grads)	400	100

■To specify the Display Format

[SHIFT][DISP] — Display: Fix Sci Nrm Eng ([F3])

[F3](Nrm) — Display: Norm_

[EXE] — Display: Norm 0.

Each time you press [SHIFT][DISP][F3](Nrm)[EXE], the display format changes between Norm 1 and Norm 2.

Important

The above specification is applied to the displayed value only. The calculator still stores the entire 13-digit mantissa and 2-digit exponent of the result in memory. If you change the display format specification while a calculation result is displayed, the display changes to show the value using your new specification.

Example **To perform 1 ÷ 200 with Norm 1, and then change to Norm 2**

[1][÷][2][0][0][EXE] — Display: 1÷200 5. E-03

Norm 1

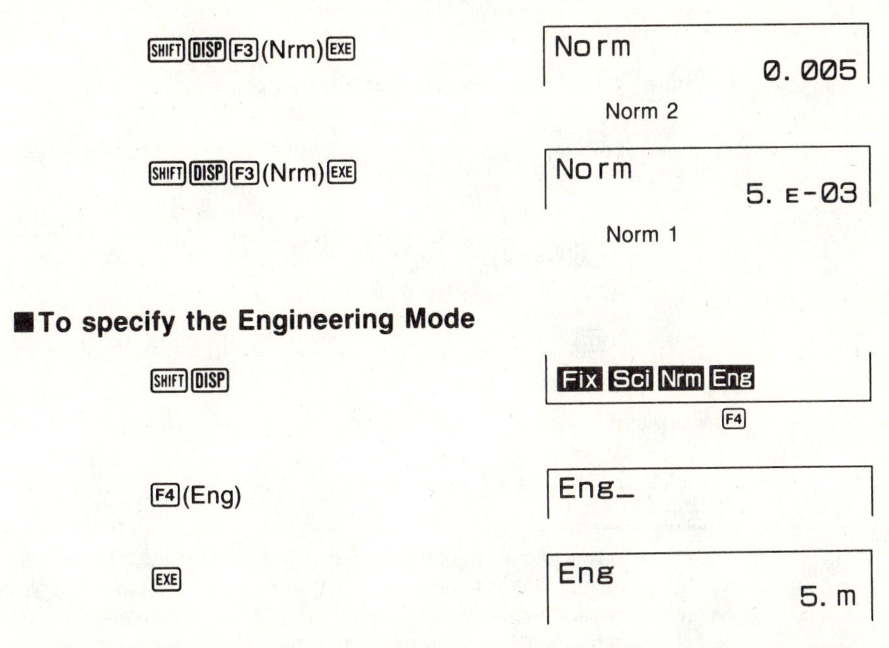

■To specify the Engineering Mode

Each time you press SHIFT DISP F4 (Eng) EXE, the unit enters or exits the Engineering Mode.

Important

The above specification is applied to the displayed value only. The calculator still stores the entire 13-digit mantissa and 2-digit exponent of the result in memory. If you change the engineering mode specification while a calculation result is displayed, the display changes to show the value using your new specification.

Example **To perform 1 ÷ 500 in Norm 1, and then change to the Engineering Mode**

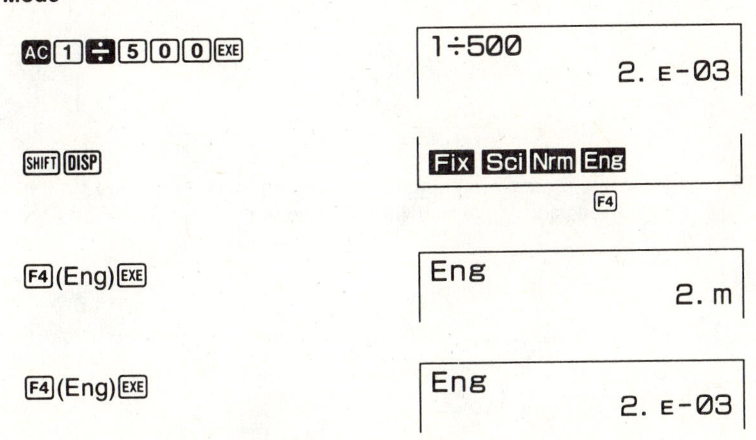

■To specify the Number of Decimal Places

Example **To set the number of decimal places to 2**

SHIFT DISP — Fix Sci Nrm Eng (F1)

F1 (Fix) 2 — Fix 2_

EXE — Fix 2 0.00

Now all displayed values will be rounded off to the nearest integer at the second decimal place.

Important

The above specification is applied to the displayed value only. The calculator still stores the entire 13-digit mantissa and 2-digit exponent of the result in memory. If you change the number of decimal places specification while a calculation result is displayed, the display changes to show the value using your new specification.

Example **To perform 100 ÷ 7 with 2 decimal places, and then change to 5 decimal places**

Note)
No matter what settings are currently being applied for the number of decimal places, pressing SHIFT DISP F3 (Nrm) EXE returns to the Norm mode (1 or 2).

■To specify the Number of Significant Digits

Example **To set the number of significant digits to 3**

SHIFT DISP — Fix Sci Nrm Eng (F2)

F2(Sci) 3 — Sci 3_

EXE — Sci 3 / 0.00E+00

Now all displayed values will be shown with 3 significant digits.

Important

The above specification is applied to the displayed value only. The calculator still stores the entire 13-digit mantissa and 2-digit exponent of the result in memory. If you change the number of significant digits specification while a calculation result is displayed, the display changes to show the value using your new specification.

Example **To perform 123 × 456 with 3 significant digits, and then change to 4 significant digits**

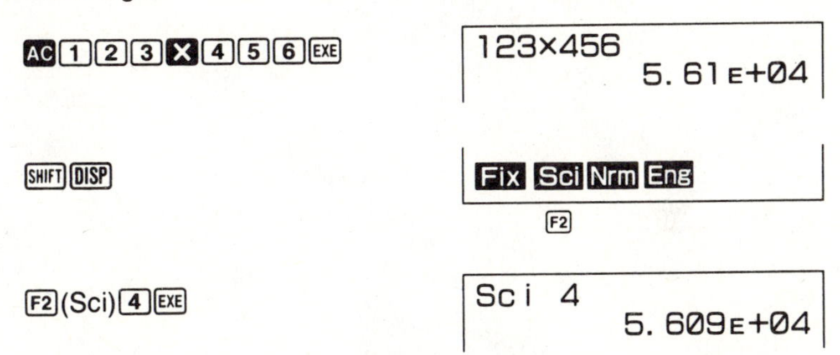

Note)
No matter what settings are currently being applied for the number of significant digits, pressing SHIFT DISP F3(Nrm) EXE returns to the Norm mode (1 or 2).

■To adjust the Contrast of the Display

MODE
◀ to make display lighter
▶ to make display darker

Important

If the display remains dim even when you adjust the contrast, you should replace batteries as soon as possible.

Basic Operation

The operations described here are fundamental calculations that you need to get started with the unit. Graphing, programming, and statistical calculations are covered in their own separate sections.

■Using the Clear Menu

The Clear Menu lets you clear either the entire memory of the unit or specific parts of the memory.

Important

- The procedures described below cannot be undone. Make sure that you do not need data any more before you delete it.
- You can call up the Clear Menu while the unit is in any mode.

● To clear the entire memory

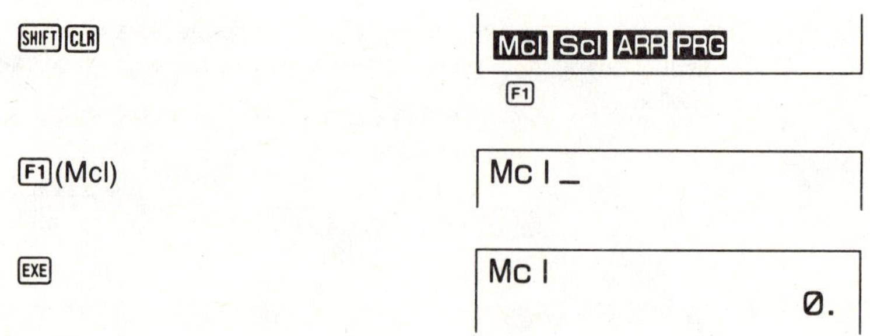

This operation clears all of the value memories, as well as any values assigned to r, θ, and variables.

● To clear statistical memories only

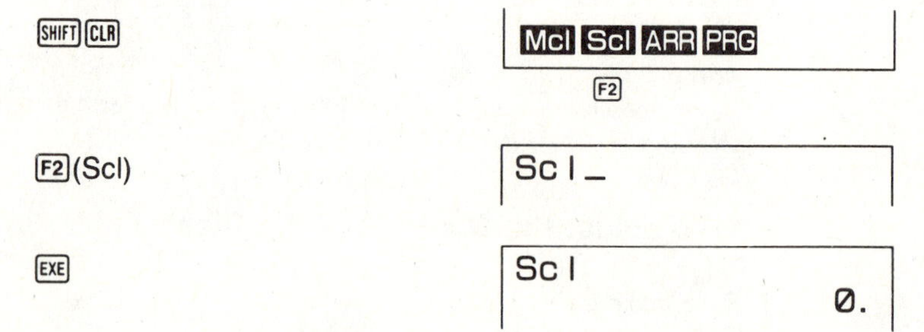

This operation clears any values assigned to Σx^2, Σx, n, Σy^2, Σy, and Σxy.

• **To clear matrix memory**

SHIFT CLR — Mcl Scl ARR PRG (F3)

F3 (ARR) — YES ERASE ARRAY NO (F1, F6)

Press F1 (YES) to clear all programs from memory or F6 (NO) (or PRE) to abort this procedure without deleting anything. This operation clears any values assigned to matrices A, B, and C.

• **To clear program memory**

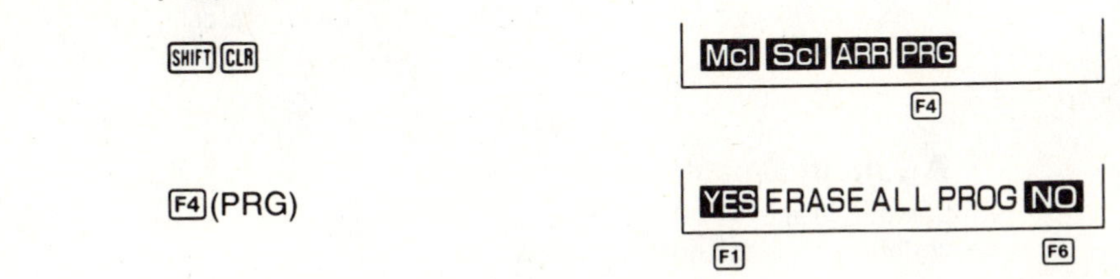

Press F1 (YES) to clear all programs from memory or F6 (NO) (or PRE) to abort this procedure without deleting anything.

■Inputting Calculations

When you are ready to input a calculation, first press AC to clear the display. Next, input your calculation formulas exactly as they are written, from left to right, and press EXE to obtain a result.

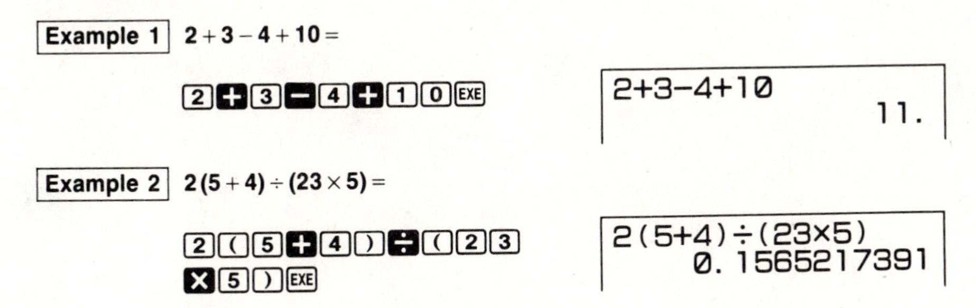

The unit uses two types of functions: Type A functions and Type B functions. With Type A functions, you press the function key after you enter a value. With Type B functions, you press the function key first and then enter a value.

Example 1 **(Type A function)**

	Example	Key Operation
Squares:	4^2	[4][x^2]

Example 2 **(Type B function)**

	Example	Key Operation
Sine:	2 sin45°	[2][sin][4][5]

• To clear an entire calculation and start again

Press the [AC] key to clear the error along with the entire calculation. Next, re-input the calculation from the beginning.

■ Editing Calculations

Use the ◀ and ▶ keys to move the cursor to the position you want to change, and then perform one of the operations described below. After you edit the calculation, you can execute it by pressing [EXE], or use ▶ to move to the end of the calculation and input more.

• To change a step

Example 1 **To change 122 to 123**

Key	Display
[1][2][2]	122_
◀	122
[3]	123_

Example 2 **To change cos60 to sin60**

Key operation	Display
[cos] [6] [0]	cos 60_
[◀] [◀] [◀]	cos 60
[sin]	sin 60

• To delete a step

Example **To change 369× ×2 to 369×2**

Key operation	Display
[3] [6] [9] [×] [×] [2]	369××2_
[◀] [◀] [DEL]	369×2

• To insert a step

Example **To change 2.36^2 to $\sin 2.36^2$**

Key operation	Display
[2] [·] [3] [6] [SHIFT] [x^2]	2. 36² _
[◀] [◀] [◀] [◀] [◀]	2. 36²
[SHIFT] [INS]	[2]. 36²
[sin]	sin [2]. 36²

•When you press [SHIFT] [INS] a space is indicated by the symbol "[ ]". The next function or value you input is inserted at the location of "[ ]". To abort the insert operation without inputting anything, move the cursor, press [SHIFT] [INS] again, or press [EXE].

• To make corrections in the original calculation

Example **14 ÷ 0 × 2.3 entered by mistake for 14 ÷ 10 × 2.3**

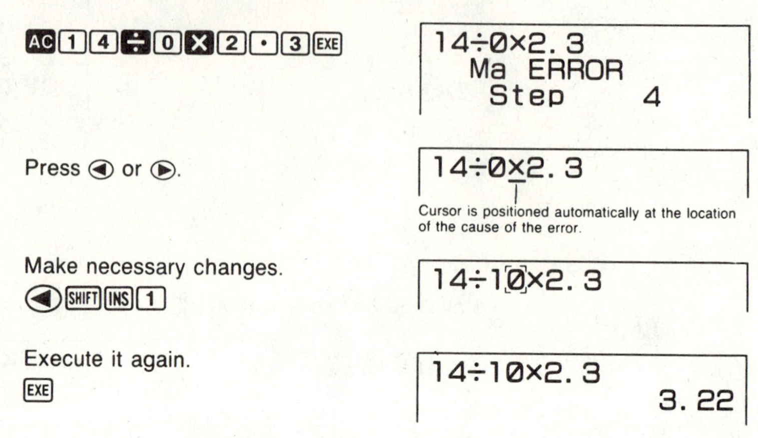

■ Answer Function

The unit's Answer Function automatically stores the last result you calculated by pressing EXE (unless the EXE key operation results in an error). The result is stored in the answer memory.

• To recall the contents of the answer memory

Ans EXE

• To use the contents of the answer memory in a calculation

Example 123 + 456 = 579
789 − 579 = 210

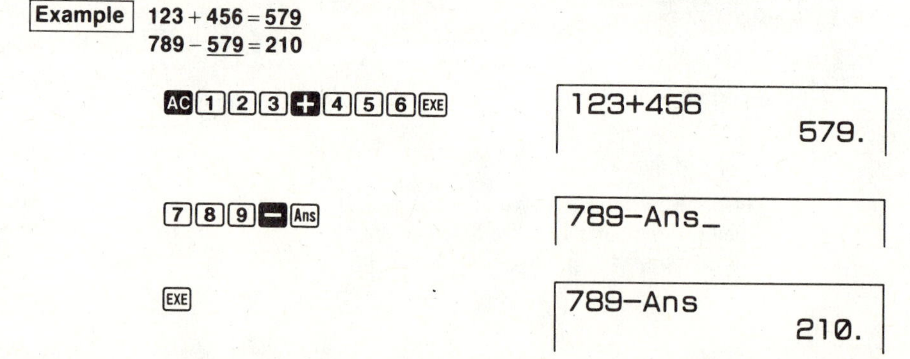

- The largest value that the answer memory can hold is one with 13 digits for the mantissa and 2 digits for the exponent.
- Answer memory contents are not cleared when you press the AC key or when you switch power off.

■Using Multistatements

Multistatements are formed by connecting a number of individual statements for sequential execution. You can use multistatements in manual calculations and in programmed calculations. There are three different ways that you can use to connect statements to form multistatements.

•Colon (:)

Statements that are connected with colons are executed from left to right, without stopping.

•Display Result Command(◢)

When execution reaches the end of a statement followed by a display result command, execution stops and the result up to that point appears on the display. You can resume execution by pressing the EXE key.

•Newline Operation

The newline operation ends the line you are currently inputting, and moves the cursor to the next line. When execution reaches the end of a line where a newline operation was performed, the unit treats the end of the line like a colon (multistatement connector).

• To use multistatements

Example $6.9 \times 123 = 848.7$
$123 \div 3.2 = 38.4375$

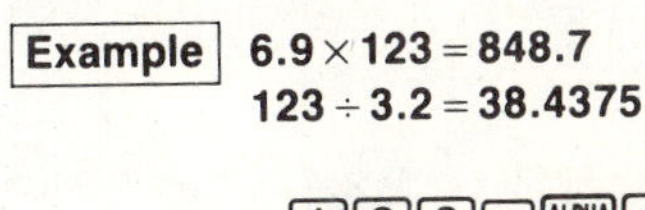

```
123→A:6. 9×A◢
A÷3. 2_
```

EXE

```
123→A:6. 9×A◢
A÷3. 2
           848. 7
         - Disp -
```

Appears on display when "◢" is used.

EXE

```
123→A:6. 9×A◢
A÷3. 2
           848. 7
          38. 4375
```

- Note that the final result of a multistatement is always displayed, regardless of whether it ends with a display result command.
- You cannot construct a multistatement in which one statement directly uses the result of the previous statement.

Example 123 × 456 : × 5

Invalid

■Multiplication Operations without a Multiplication Sign

You can omit the multiplication sign (×) in any of the following operations.

- Before the type B functions and coordinate transformation functions:

Example **2sin30, 10log1.2, 2√3, 2pol(5, 12), etc.**

- Before constants, variable names, value memory names

Example **2π, 2AB, 3Ans, etc.**

- Before an open parenthesis

Example **3(5 + 6), (A + 1)(B − 1), etc.**

■Performing Continuous Calculations

The unit lets you use the result of one calculation as one of the arguments in the next calculation. The precision of such calculations is 10 digits (for the mantissa).

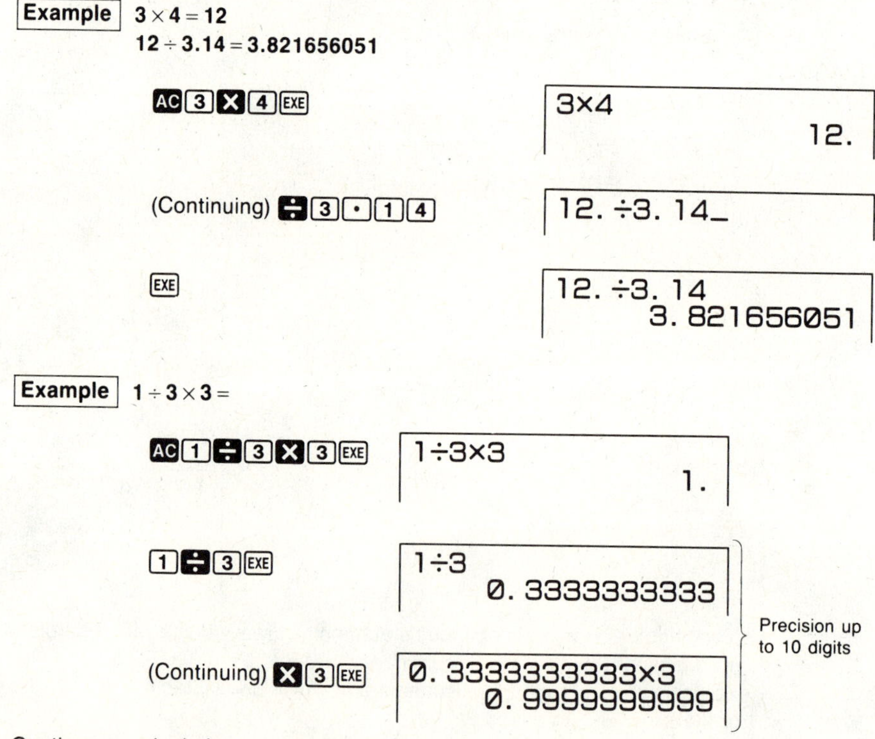

Continuous calculations can also be used with Type A functions.

Example $78 \div 6 = 13$
$13^2 = 169$

AC 7 8 ÷ 6 EXE — Display: 78÷6 / 13.

(Continuing) SHIFT x^2 — Display: 13. ²_

EXE — Display: 13. ² / 169.

■ Using the Replay Function

The Replay Function automatically stores the last calculation performed in replay memory. You can recall the contents of the replay memory by pressing ◀ or ▶. If you press ▶, the calculation appears with the cursor at the beginning. Pressing ◀ causes the calculation to appear with the cursor at the end. You can make changes in the calculation as you wish and then execute it again.

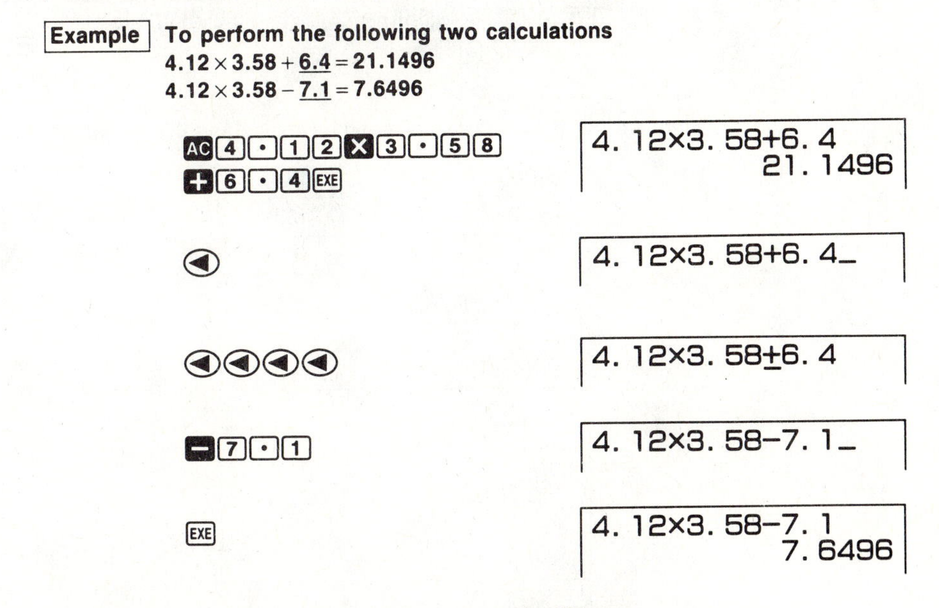

- The maximum capacity of the replay memory is 127 steps.
- The contents of the replay memory are retained even if you press AC or switch power off.

Example

[AC] [1] [2] [3] [×] [4] [5] [6] [EXE]

```
123×456
          56088.
```

[AC]

```
_
```

[◀]

```
123×456_
```

*The contents of the replay memory are cleared whenever you change from one menu to another.

■Engineering Symbols

You can call up this menu to select engineering symbols for use in calculations.

● To use engineering symbols in calculations

Example **1000 m × 5 k**

[1] [0] [0] [0] [SHIFT] [ENG SYM] [F1] (m) [×] [5] [F6] (▽) [F1] (k) [EXE]

```
1000m×5k
           5000.
```

The following is a list of available engineering symbols and their meanings.

[SHIFT] [ENG SYM] displays:

[F1] (m)	milli	10^{-3}
[F2] (μ)	micro	10^{-6}
[F3] (n)	nano	10^{-9}
[F4] (p)	pico	10^{-12}
[F5] (f)	femto	10^{-15}
[F6] (▽) Next menu		

[F6] displays:

[F1] (k)	kilo	10^{3}
[F2] (M)	mega	10^{6}
[F3] (G)	giga	10^{9}
[F4] (T)	tera	10^{12}
[F5] (P)	peta	10^{15}
[F6] (E)	exa	10^{18}

■Scientific Functions

There are 4 scientific function menus: a Hyperbolic Function Menu, a Probability Function Menu, a Numeric Function Menu, and a Sexagesimal Function Menu.

● To call up the Scientific Function Menu

SHIFT MATH — HYP PRB NUM DMS

● To use the Hyperbolic Function Menu

SHIFT MATH — HYP PRB NUM DMS

F1

F1(HYP) — snh csh tnh snh^{-1} csh^{-1} tnh^{-1}

F1 F2 F3 F4 F5 F6

Press the function key below the hyperbolic function you want to input.

- F1(snh) .... hyperbolic sine
- F2(csh) .... hyperbolic cosine
- F3(tnh) .... hyperbolic tangent
- F4(snh^{-1}) .... inverse hyperbolic sine
- F5(csh^{-1}) .... inverse hyperbolic cosine
- F6(tnh^{-1}) .... inverse hyperbolic tangent

Press PRE to backtrack to the Scientific Function Menu.

● To use the Probability Function Menu

SHIFT MATH — HYP PRB NUM DMS

F2

F2(PRB) — $x!$ nPr nCr Rn#

F1 F2 F3 F4

Press the function key below the probability function you want to input.

- F1($x!$) .... factorial of x
- F2(nPr) .... permutation
- F3(nCr) .... combination
- F4(Rn#) .... random number generation

Press PRE to backtrack to the Scientific Function Menu.

• To use the Numeric Function Menu

[SHIFT][MATH] — HYP PRB NUM DMS
F3

[F3](NUM) — Abs Int Frc Rnd
F1 F2 F3 F4

Press the function key below the numeric function you want to input.

[F1](Abs) .... absolute value
[F2](Int) .... integer extraction
[F3](Frc) .... fraction extraction
[F4](Rnd) .... rounding

Press [PRE] to backtrack to the Scientific Function Menu.

• To use the Sexagesimal Function Menu

[SHIFT][MATH] — HYP PRB NUM DMS
F4

[F4](DMS) — ° ' " ←° ' "
F1 F2

Press the function key below the sexagesimal function you want to input.

[F1](° ' ") .... For input of hours, minutes and seconds, or degrees, minutes and seconds as sexagesimal values
[F2](←° ' ") .... For input of hours, minutes and seconds, or degrees, minutes and seconds as decimal values

Press [PRE] to backtrack to the Scientific Function Menu.

■Value Memories

The unit comes with 28 value memories as standard (which can be expanded up to 548). You can use value memories to store values to be used inside of calculations. Value memories are identified by single-letter names, which are made up of the 26 letters of the alphabet, plus r and θ. The maximum size of values that you can assign to value memories is 13 digits for the mantissa and 2 digits for the exponent. Value memory contents are retained even when you switch power off.

Important

• Some value memories are used by the unit for certain types of calculations. Note the following.

Type of Calculation	Value Memories Used
Single-Variable Statistics (non-storage)	U, V, W
Paired-Variable Statistics (non-storage)	P, Q, R, U, V, W
Integration	K, L, M, N
Coordinate Conversion	I, J

You cannot assign values to these value memories while the above calculations are being performed. You should also clear the value memories before starting the above operations. Be especially careful during programmed calculations to avoid problems caused by values mistakenly assigned to memories that are used by the calculator.

• To assign a value to a value memory

Example To assign 123 to value memory A

[1] [2] [3] [→] [ALPHA] [A] [EXE]

```
123→A
                123.
```

Example To add 456 to value memory A and store the result in value memory B

[ALPHA] [A] [+] [4] [5] [6] [→] [ALPHA] [B] [EXE]

```
A+456→B
                579.
```

• To store the result of an operation to a value memory

Example To store the result of log2 to value memory S

[log] [2] [→] [ALPHA] [S] [EXE]

```
log 2→S
      0.3010299957
```

• To display the contents of a value memory

Example **To display the contents of value memory A**

[ALPHA][A][EXE]

```
A
                123.
```

• To clear a value memory

Example **To clear value memory A**

[0][→][ALPHA][A][EXE]

```
0→A
                  0.
```

• To clear all value memory contents

[SHIFT][CLR][F1](Mcl)[EXE]

```
Mcl
                  0.
```

■Increasing the Number of Value Memories

Though 28 value memories are provided as standard, you can configure the memory of the unit to increase the number of value memories and decrease the amount of program memory. Each additional value memory decreases the number of program memory steps by 8.

Number of Value Memories	28	29	30	31	·······	548
Number of Program Memory Steps	4164	4156	4148	4140	·······	4

The maximum number of value memories possible is 548 (an increase of 520).

Important

- You may not be able to increase the number of value memories to the level you want if the memory already contains programs, matrices, function memory contents, or statistical data. If there is not enough unused memory available to increase to the number you specify, an error message will appear on the display.
- The [SHIFT][Defm] specification can also be included within a program.

● To increase the number of value memories

Example **To increase the number of value memories by 30 (for a total of 28 + 30 = 58)**

[SHIFT] [Defm] [3] [0] [EXE]

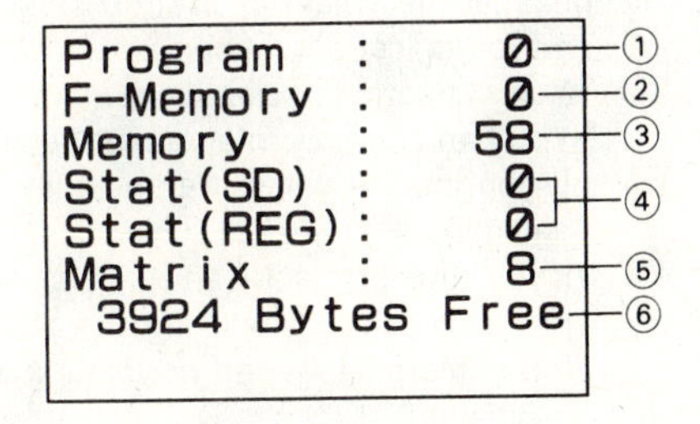

① Number of Program Steps Used
② Number of Function Memory Steps Used
③ Number of Value Memories Available
④ Number of Statistical Memory Steps Used
⑤ Number of Matrix Memory Steps Used
⑥ Number of Unused Program Steps Remaining

● To check the current memory status

[SHIFT] [Defm] [EXE]

● To initialize the number of value memories

[SHIFT] [Defm] [0] [EXE]

```
Program  :    0
F-Memory :    0
Memory   :   28
Stat(SD) :    0
Stat(REG):    0
Matrix   :    8
 4164 Bytes Free
```

■ About memory names

You can use the additional memories you create from program memory just as you use the original 28. The names of the additional memories are Z[1], Z[2], Z[3], etc. If you increase the number of value memories by 5, you can access the original 28 memories, plus memories Z[1] through Z[5].

Using the BASE-N Mode

You can use the BASE-N Mode to perform calculations with binary, octal, decimal and hexadecimal values. You should also use this mode to convert between number systems and for logical operations.

- You cannot use scientific functions in the BASE-N Mode.
- You can use only integers in the BASE-N Mode, so fractional values are not allowed. If you input a value that includes a decimal part, the unit automatically cuts off the decimal.
- If you attempt to enter a value that is invalid in the number system (binary, octal, decimal, hexadecimal) you are using, the calculator displays an error message. The following show the numerals that can be used in each number system.

  Binary: 0, 1
  Octal: 0, 1, 2, 3, 4, 5, 6, 7
  Decimal: 0, 1, 2, 3, 4, 5, 6, 7, 8, 9
  Hexadecimal: 0, 1, 2, 3, 4, 5, 6, 7, 8, 9, A, B, C, D, E, F
- The alphabetic characters used in the hexadecimal number appear differently on the display to distinguish them from text characters.

  Normal Text: A, B, C, D, E, F
  Hexadecimal Values: **A**, **B**, **C**, **D**, **E**, **F**
- Negative binary, octal, and hexadecimal values are produced using the two's complement of the original value.
- The following are the display capacities for each of the number systems.

Number System	Display Capacity
Binary	16 digits
Octal	11 digits
Decimal	10 digits
Hexadecimal	8 digits

- The following are the calculation capacities for each of the number systems.

  Calculation Ranges in BASE-N Mode

  Binary Values
    Negative : $1000000000000000 \leqq x \leqq 1111111111111111$
    Positive : $0 \leqq x \leqq 111111111111111$

  Octal Values
    Negative : $20000000000 \leqq x \leqq 37777777777$
    Positive : $0 \leqq x \leqq 17777777777$

  Decimal Values
    Negative : $-2147483648 \leqq x \leqq -1$
    Positive : $0 \leqq x \leqq 2147483647$

  Hexadecimal Values
    Negative : $80000000 \leqq x \leqq FFFFFFFF$
    Positive : $0 \leqq x \leqq 7FFFFFFF$

• To enter the BASE-N Mode

[MODE][-]

Main BASE-N Mode screen

```
RUN / BASE-N
        DEC

Dec Hex Bin Oct d~o LOG
F1  F2  F3  F4  F5  F6
```

• To set the default BASE-N Mode number system

Example **To set the default BASE-N Mode number system to decimal**

[F1](Dec)[EXE]

```
Dec
                0
```

The following are the number systems that are available.

- [F1](Dec) .... decimal
- [F2](Hex) .... hexadecimal
- [F3](Bin) .... binary
- [F4](Oct) .... octal

• To convert a displayed value from one number system to another

Example **To convert 1,038$_D$ (default number system) to its hexadecimal value**

[1][0][3][8][EXE]

```
1038
            1038
```

[F2](Hex)[EXE]

```
Hex
        0000040E
```

• To input values of mixed number systems

Example **To input $1{,}038_D + 25C_H + 11011_B + 23_O$, when the default number system is decimal**

[F1](Dec)[EXE]
[1][0][3][8][+][F5](d ~ o)[F2](h)
[2][5][C][+][F3](b)[1][1][0][1][1]
[+][F4](o)[2][3][EXE]

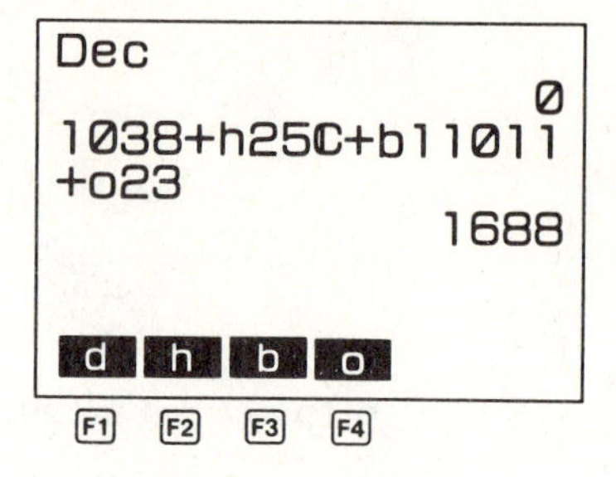

The following are the types of values that can be specified in the above menu.

[F1](d) .... decimal value
[F2](h) .... hexadecimal value
[F3](b) .... binary value
[F4](o) .... octal value

Press [PRE] to backtrack to the main BASE-N Mode screen.

• To input logical operations

Example **To input and execute "120_{16} and AD_{16}"**

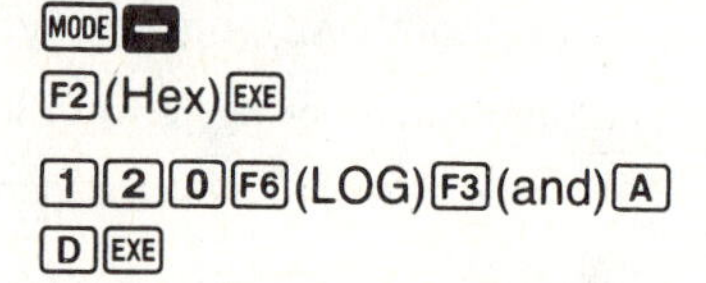

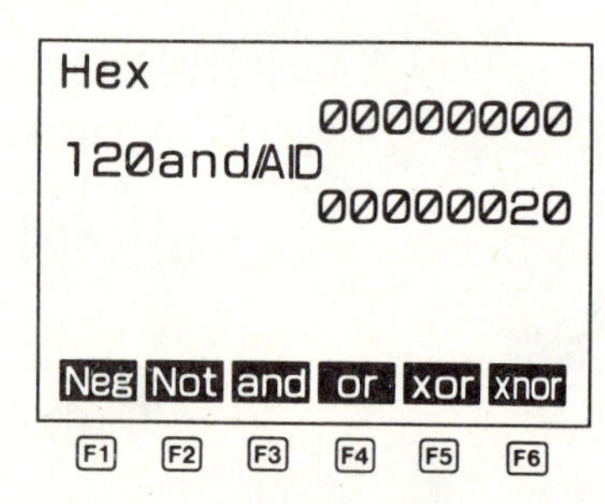

The following are the logical operations that can be input from the above menu.

[F1](Neg) .... negation
[F2](Not) .... NOT
[F3](and) .... AND
[F4](or) .... OR
[F5](xor) .... XOR
[F6](xnor) .... XNOR

Press [PRE] to backtrack to the main BASE-N Mode screen.

Using the Matrix Mode

■ About Matrices

This unit's matrix operations use 3 matrices, named A, B, C. The following table shows how each matrix is used.

Matrix Name	A	B	C
Addition/Subtraction/ Multiplication/Division	○	○	Result
Scalar Product	○	○	Result
Transposition Matrix	○	○	Result
Determinant	○	○	Not used
Inverse Matrix	○	○	Result
Matrix Exchange	Exchange of A and B		Not used
Matrix C Copy	Destination		Origin
Matrix Dimension	9×9 maximum		

• To enter the Matrix Mode

MODE 0

Main Matrix Mode screen

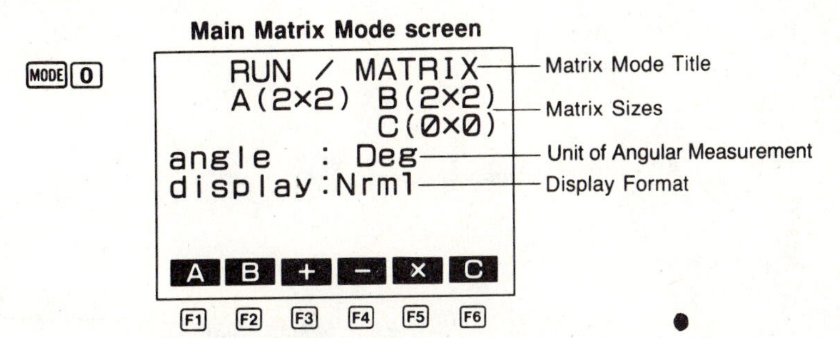

The following are the operations that are available from this menu. Press the function key below the operation you want to perform.

F1(A) .... Displays matrix A contents
F2(B) .... Displays matrix B contents
F3(+) .... Adds matrix A and matrix B
F4(−) .... Subtracts matrix B from matrix A
F5(×) .... Multiplies matrix A and matrix B
F6(C) .... Displays matrix C contents

Important

Many of the matrix operations described in this manual are performed using matrix A in examples. Note that the same operations can be used with matrix B.

● **To clear matrix memory**

SHIFT CLR F3 (ARR)

Press F1 (YES) to clear matrix memory or F6 (NO) (or PRE) to abort the operation without clearing anything.
You should clear matrix memory if you want to perform any non-matrix calculations that use memories. Note that the above operation is not required if you have specified a new matrix size, because the size specification automatically clears matrix memory.

● **To specify matrix size**

Example **To specify a size of 3 × 3 for matrix A**

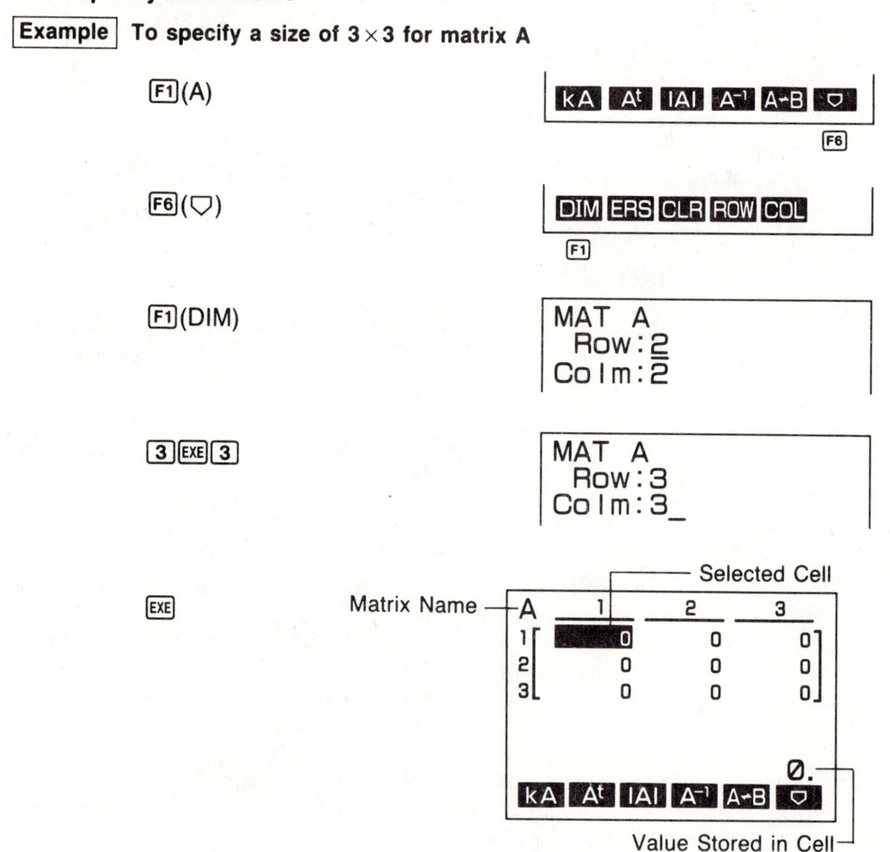

• To input matrix data

Example **To input the following data into matrix A (3×4)**

$$\begin{pmatrix} 1 & 0 & 3 & 4 \\ 2 & 1 & 0 & 1 \\ 3 & 1 & -2 & -3 \end{pmatrix}$$

Input each value and press [EXE].

[1] [EXE] [0] [EXE] [3] [EXE] [4] [EXE]
[2] [EXE] [1] [EXE] [0] [EXE] [1] [EXE]
[3] [EXE] [1] [EXE] [−] [2] [EXE] [−] [3] [EXE]

• After you finish inputting the data, you can return to the main Matrix Mode display by pressing [PRE].

• To move around a matrix

You can move around the matrix using the cursor keys.

[▲] **Moves up.**
[▼] **Moves down.**
[◀] **Moves left.**
If the pointer is at the far left of a row and there is another row above, pressing this key scrolls to the line above, with the pointer at the far right of the line.
[▶] **Moves right.**
If the pointer is at the far right of a row and there is another row below, pressing this key scrolls to the line below, with the pointer at the far left of the line.
• Holding down any of the cursor keys performs the corresponding operation at high speed.

• Capacity of each cell

• Only 5 rows and 3 columns of a matrix can be shown on the display. The cursor key operations cause the screen to scroll in order to accommodate larger matrices.
• The capacity for each cell is 6 digits for positive values and 5 digits for negative values.
• Exponential values are cut off to one significant digit.

■Performing Matrix Arithmetic Operations

You can use matrix A and matrix B contents in addition, subtraction and multiplication operations. The examples of these operations presented here are based on the following 2 matrices.

Matrix A $\begin{pmatrix} 1 & 1 \\ 2 & 1 \end{pmatrix}$ **Matrix B** $\begin{pmatrix} 2 & 3 \\ 2 & 1 \end{pmatrix}$

Create these matrices in memory using the following procedure.

● To input matrix A data

[MODE][0][F1](A)[F6](▽)[F1](DIM)
[2][EXE][2][EXE]

[1][EXE][1][EXE]
[2][EXE][1][EXE]

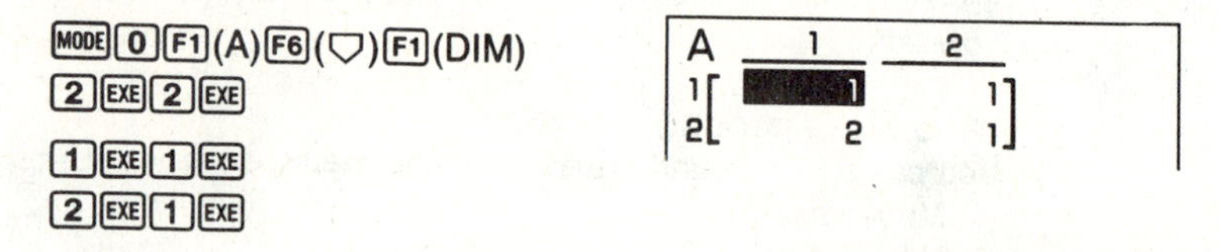

● To input matrix B data

[PRE][F2](B)[F6](▽)[F1](DIM)
[2][EXE][2][EXE]

[2][EXE][3][EXE]
[2][EXE][1][EXE]

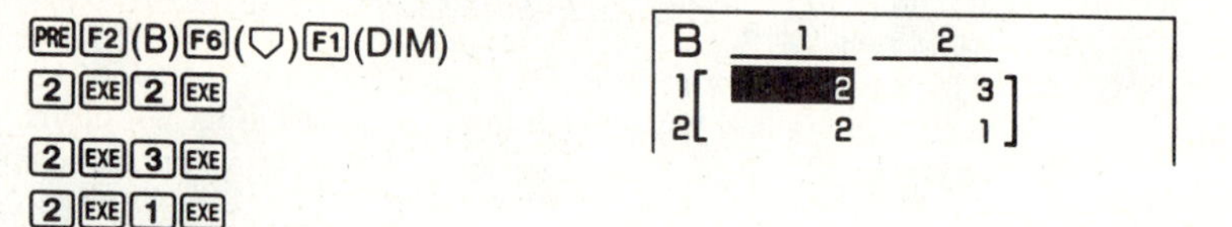

● To add matrix A and matrix B

[PRE]

[F3](+)

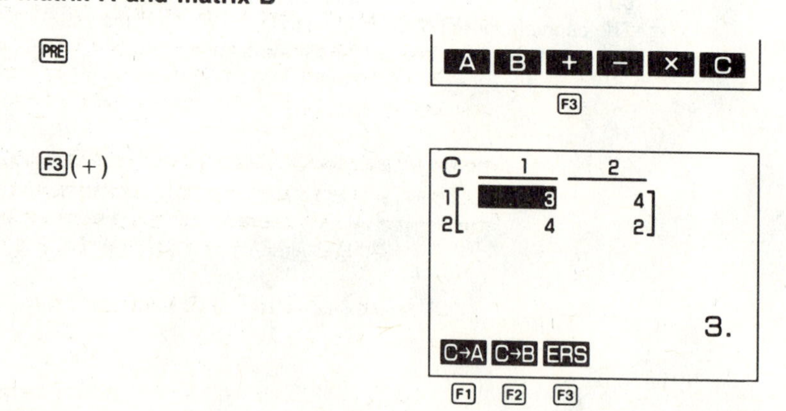

•Matrix C appears, showing the sum of the values in the cells of matrix A and matrix B.

•The following are the operations that are available from the function display at the bottom of the screen. Press the function key below the operation you want to perform.

F1(C→A) .... Transfers matrix C contents to matrix A (deleting matrix A contents)

F2(C→B) .... Transfers matrix C contents to matrix B (deleting matrix B contents)

F3(ERS) .... Deletes the matrix

Important

•Matrix A and matrix B can be added only if the dimensions of the matrices are identical. Different dimensions produce a "Dim ERROR" when you try to add the matrices.

• To subtract matrix B from matrix A

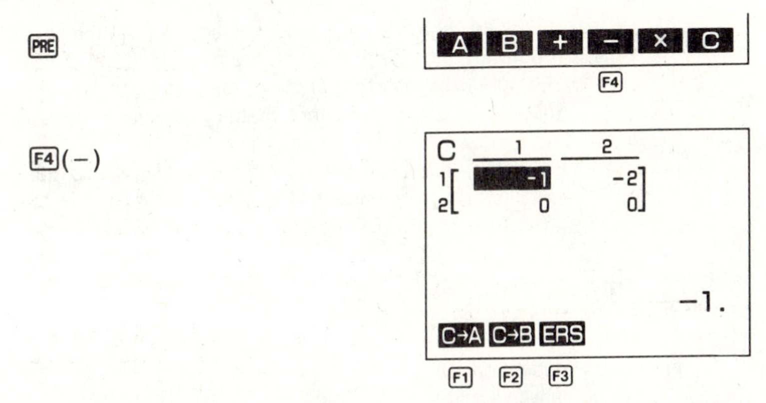

•Matrix C appears, showing the difference of the values in the cells of matrix A and matrix B.

•The following are the operations that are available from the function display at the bottom of the screen. Press the function key below the operation you want to perform.

F1(C→A) .... Transfers matrix C contents to matrix A (deleting matrix A contents)

F2(C→B) .... Transfers matrix C contents to matrix B (deleting matrix B contents)

F3(ERS) .... Deletes the matrix

Important

•Matrix A and matrix B can be subtracted only if the dimensions of the matrices are identical. Different dimensions produce a "Dim ERROR" when you try to subtract the matrices.

•You cannot subtract matrix A from matrix B. To accomplish the equivalent result though, you can exchange the contents of matrix A and matrix B.

• To multiply matrix A by matrix B

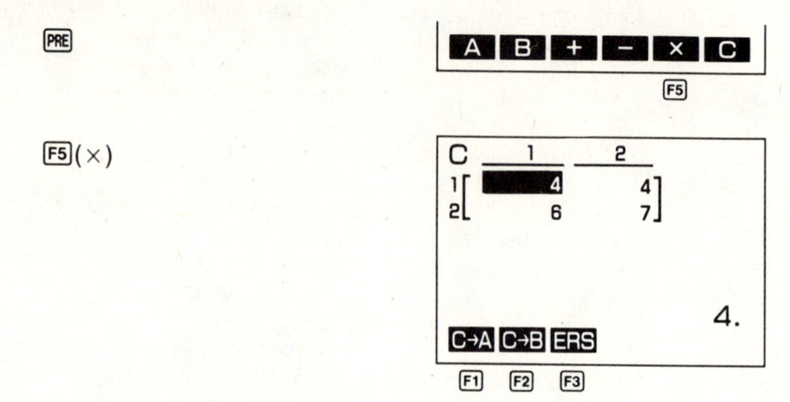

• Matrix C appears, showing the product of the values in the cells of matrix A and matrix B.
• The following are the operations that are available from the function display at the bottom of the screen. Press the function key below the operation you want to perform.

F1(C→A) .... Transfers matrix C contents to matrix A (deleting matrix A contents)
F2(C→B) .... Transfers matrix C contents to matrix B (deleting matrix B contents)
F3(ERS) .... Deletes the matrix

Important

• Matrix A and matrix B can be multiplied only if they are of identical size (but not necessarily of identical dimensions). If matrix A is 3×2, for example, it can be used for multiplication with a matrix B that is $2 \times n$ ($n = 1 \sim 9$). Different sizes produce a "Dim ERROR" when you try to multiply the matrices.
• You cannot multiply matrix B by matrix A. To accomplish the equivalent result though, you can exchange the contents of matrix A and matrix B.

• To calculate an inverse matrix

Example **To calculate the inverse matrix of the following data**

Matrix A

$$\begin{pmatrix} 1 & 2 \\ 3 & 4 \end{pmatrix}$$

[F1](A)

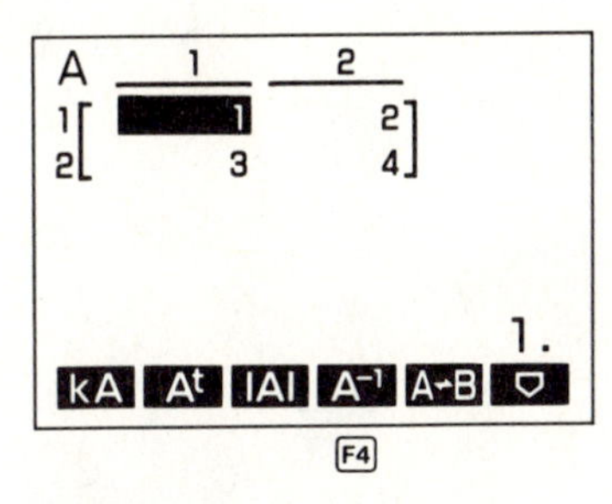

[F4](A^{-1})

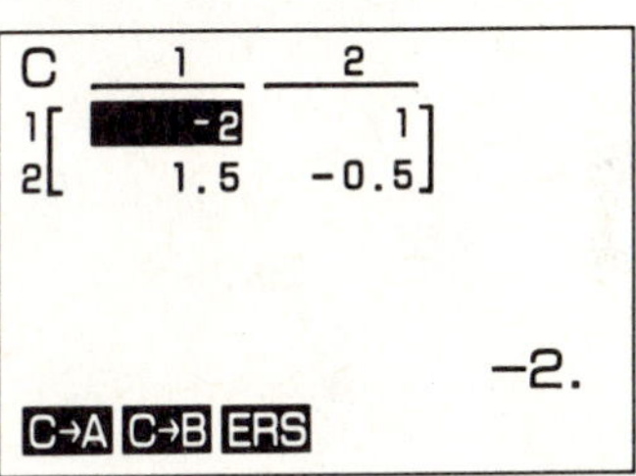

- This operation calculates the inverse of square matrix A or B and stores the results in matrix C.
- The dimension of matrix C is the same as matrix A or B.
- There is no inverse matrix when $ad - bc = 0$ (when the matrix equals 0). In such a case, the above operation produces a "Ma ERROR".
- Note that the inverse matrix can be calculated for square matrices (same number of rows and columns) only. A "Dim ERROR" occurs when this operation is attempted with a matrix that is not a square matrix.
- Matrix A^{-1} (which is the inverse of matrix A) satisfies the following conditions.

$$\mathbf{AA-1=E=}\begin{pmatrix} 1 & 0 \\ 0 & 1 \end{pmatrix}$$

- The following is applied to the inverse matrix (A^{-1}) of 2×2 square matrix A.

$$\mathbf{A}=\begin{pmatrix} a & b \\ c & d \end{pmatrix}$$

Therefore, $\mathbf{A}^{-1}=\frac{1}{ad-bc}\begin{pmatrix} d & -b \\ -c & a \end{pmatrix}$ when $ad-bc \neq 0$

• To exchange matrix A and matrix B contents

Example **To exchange the contents of matrix A and matrix B when they originally contain the following data**

Matrix A $\begin{pmatrix} 1 & 2 \\ 3 & 4 \end{pmatrix}$ **Matrix B** $\begin{pmatrix} -1 & 2 \\ 3 & 4 \end{pmatrix}$

F1 (A)

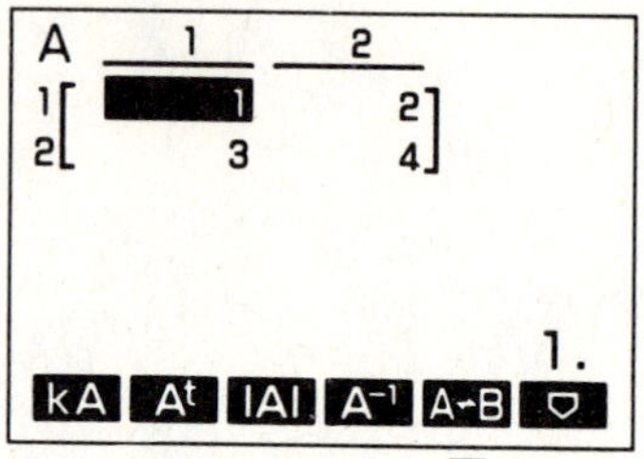

F5 (A↔B)

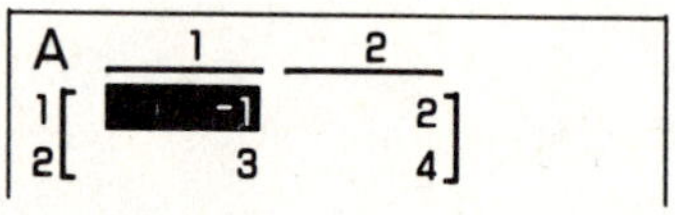

•This operation exchanges the contents of matrix A and matrix B.

Using the Function Memory

You can store up to six functions in memory for instant recall when you need them. Function memory can be used in any mode except the BASE-N Mode.

● To display the Function Memory Menu

SHIFT F·MEM

• The following are the operations that are available from the function display at the bottom of the screen. Press the function key below the operation you want to perform.

- F1(STO) .... Stores functions
- F2(RCL) .... Recalls functions
- F3(fn) .... Specifies input as a function.
- F4(LIST) .... Displays a list of stored functions

● To store a function

Example **To store the function (A + B) (A − B) as function memory number 3.**

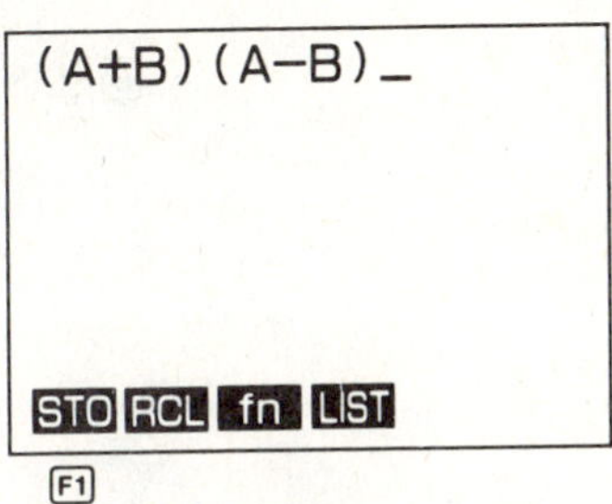

F1

F1(STO)

3

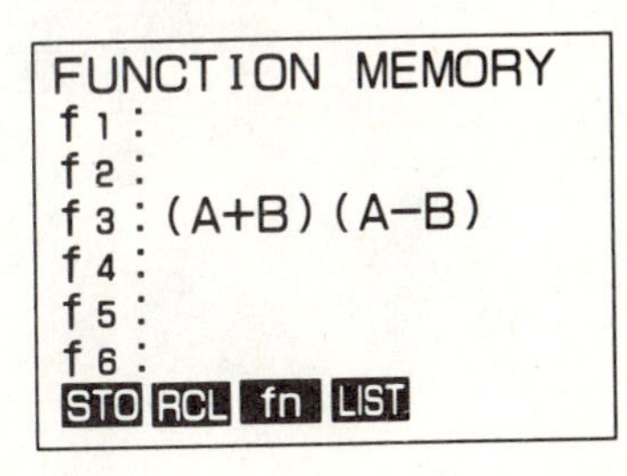

• If the function memory number you assign a function to already contains a function, the previous function is replaced with the new one.

• To recall a function

Example **To recall function memory number 3**

SHIFT F MEM → STO RCL fn LIST

F2

F2 (RCL) → STO RCL fn LIST

3 → (A+B)(A−B)_

•The recalled function appears at the current location of the cursor on the display.

• To display a list of available functions

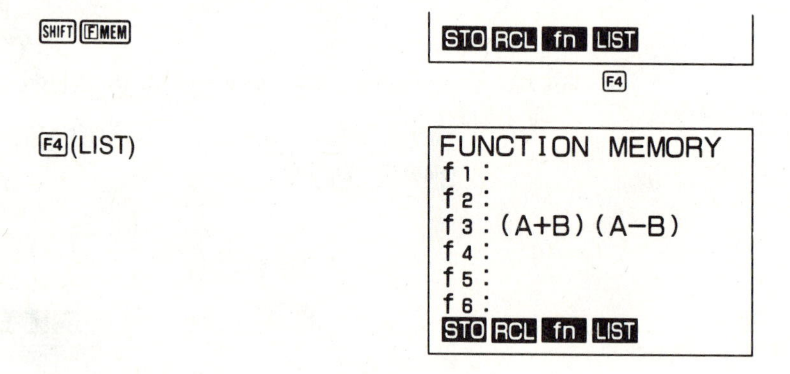

• To delete a function

Example **To delete function memory number 3**

SHIFT F MEM → STO RCL fn LIST

AC

F1

F1 (STO) → STO RCL fn LIST

3 →

FUNCTION MEMORY
f1 :
f2 :
f3 :
f4 :
f5 :
f6 :
STO RCL fn LIST

•Executing the store operation while the display is blank deletes the function for the Function Memory you specify.

Sharp EL–9200 and EL–9300

Calculating and Graphing with the Sharp EL-9200 and EL-9300

The Sharp EL-9200 and EL-9300 offer a mathematics student an extremely wide range of functionality in a user-friendly format. Both calculators feature almost identical software which includes; rectangular, polar and parametric graphing, programming, statistics, statistical graphing, matrices, complex number calculations, and numerical calculus. The calculators feature a large high contrast screen and a hard case for protection when not in use. The EL-9300 also features a powerful Solver function, 32K of RAM, a communications port and a backup battery. These brief instructions apply to either the EL-9200 or EL-9300 except for the Solver operation which applies only to the EL-9300.

Both calculators have a built in self demonstration that can be viewed by holding down the [ENTER] key and pressing [ON] when the calculator is turned off.

.10 General Information

.11 Keyboard layout

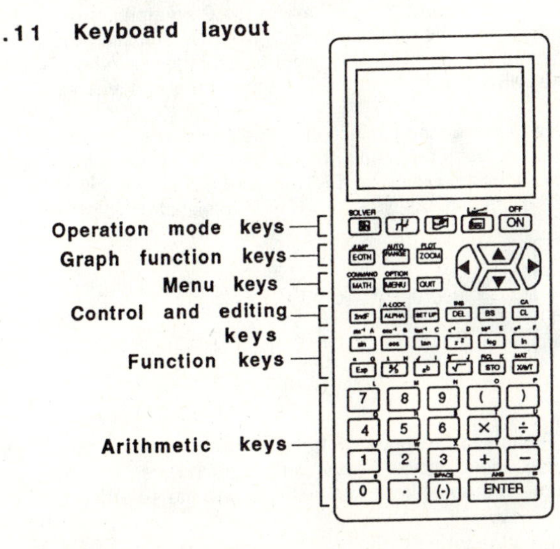

The keys on the calculator are grouped and colored for easier use. The major key groups are shown on the left. Every key has a primary function that is printed on the key in white. Most keys have second functions which are printed above the key in yellow. These keys are accessed by pressing the yellow [2ndF] key first followed by the key desired. For example $\sin^{-1}$ can be accessed by pressing [2ndF] [$\sin^{-1}$]. Many keys also have a third function printed in blue above and to the right of the key. These letters are accessed by pressing [ALPHA] followed by the letter. The calculator can be locked into Alpha mode by pressing [2ndF] [A-LOCK]. Pressing [ALPHA] a second time clears the Alpha Lock. Indicators at the top of the screen show if the Alpha or 2ndF keys have been pressed.

The top row of keys, the **Operation Mode keys**, are used to choose the mode you want to work in. From left to right they are; Calculation mode, Graph mode, Programming mode, and Statistics mode. There are two additional modes available as second functions that are; Solver mode (EL-9300 only) and Statistical Graph mode. To go to any mode simply press the key for the mode you want. The calculator will instantly go into the selected mode.

The second row of keys are the **Graph Function keys**. These keys are for various operations in the Graphing modes.

The third row of keys are the **Menu keys**. These keys call up menus with functions and operations not shown on the keyboard. Operation of the menus is described in section .12.

This information is from the *Sharp EL–9200 and EL–9300 Graphic Scientific Calculators Owner's Manual and Solutions Handbook* published by Sharp and is used with permission.

Row four contains the **Control and Editing keys**. [SET UP] and [2ndF] [INS] control the operation of the calculator. [DEL], [BS] (Backspace), [CL] and [2ndF] [CA] allow you to edit expressions and equations entered in the calculator.

The **Function keys** and **Arithmetic keys** are the same as on any scientific or graphing calculator. The popular [X] key is located to the right side just above the parenthesis keys. This key gives an X, θ or T depending on the graph coordinate system selected in Set Up.

.12 Menu Operation

Many operations and functions on the Sharp calculators are accessed through a two level menu system. All of the menus work in the same manner.

The two menus used most often are the [MATH] and [MENU] menus. The **Math** menu contains all of the additional *functions* available for the current mode. Functions which are not available will have a row of dots instead of the function. The **Menu** menu contains all of the additional *operations* available for the current mode. This menu will change completely depending of the current mode. The **Command** menu is used only in the Programming mode. It contains all of the *programming commands*. The **Option** menu is used for adjusting the contrast, memory management and communications (EL-9300 only) to a printer, another EL-9300, or a cassette player. The [QUIT] key always exits from the current menu and saves all of the selections.

There are several ways to make menu selections. The easiest way is by using the 4 arrow keys. As the user scrolls down through the menu, the sub-menu changes. When the correct sub-menu is displayed, press the right arrow key or [ENTER] and the cursor will move into the sub-menu. Use the up and down cursor keys and [ENTER] to select choices from the sub-menu.

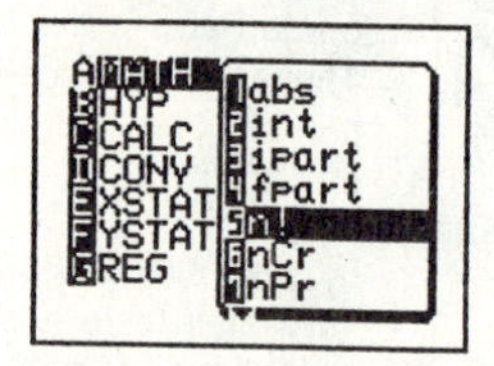

You can also use the Alpha and Number keys to make quick menu selections. Press the letter key corresponding to the main menu choice. These is no need to press [ALPHA]. When the correct sub-menu is showing press the number of your choice and the selection will be made.

.13 Equation Editor

The Equation Editor allows equations and expressions to be entered, edited, and viewed in the exact format in which they appear on paper. This makes expressions much easier to see and understand. The Equation Editor is the default editor on both the EL-9200 and EL-9300 and is used in the Calculator, Graphing and Solver modes of the calculator. A traditional 'One Line' editor is also available which will make the calculator operate in the same manner as models from Texas Instruments and Casio. This can be turned on by pressing [SET UP] [F] [2]. The rest of these instructions will assume that the calculator is set to the Equation Editor.

$\frac{2^6(1-2^6)}{6*6!\pi^6}=$

-0.000970817

$\frac{2^8(1-2^8)}{8*8!\pi}=$

Expressions are entered as they would be on any other graphing or programmable calculator except that the expression appears in two dimensions. Use the right arrow key to move out from pending functions such as exponents, roots or out from under a fraction. All 4 arrow keys can be used to edit the expression and [2ndF] [<] or [>] brings you to the beginning or end of the expression. To compute the answer press [ENTER]. It is not necessary to be at the end of the expression when [ENTER] is pressed.

.30 Graphing

.31 Y1 - Y4

You can graph and trace up to 4 equations at one time. These four equations are labeled Y1 through Y4. When you first enter the Graph mode by pressing the Graph key (in the top row of keys), 'Y1=' will appear on the screen. Entering an expression here is done in the same manner as in the Calculation mode. The horizontal axis is always represented by X in the Graph mode. Use the [X/θ/T] key to quickly enter a X in the equation. Press [ENTER] to store the expression, and the calculator will move to Y2. You can use [2ndF] followed by the up or down arrow keys to move between the 4 expressions. After Y4 there is a Fill screen which allows you to shade above and below Y1 through Y4. To return to the equation screen from any other graph screen press the [EQTN] key.

For Example: Graph $\frac{\sin X}{X-1}$ and $\frac{1}{X-1}$:

Press the Graph Key to start the Graph mode. The screen will show 'Y1=' (press [CL] if any other equation is showing). Enter the first equation.

[a/b] [sin] [X/θ/T] [>] [X/θ/T] [-] [1] [ENTER]

Y1= $\frac{\sin X}{X-1}$

The first equation is stored and the screen shows 'Y2='. Enter the second equation:

[1] [a/b] [X/θ/T] [-] [1] [ENTER]

If you wanted to graph more equations, you could enter them as Y3 and Y4.

.32 RANGE and AUTO RANGE

Before drawing the graph, you must set the range (or the viewing rectangle). Press the [RANGE] key and the current settings for the X Range are shown. Press the left arrow key and the Y Range values are shown. You can enter new values for the range, or press the [MENU] key and select from 19 default ranges available. Since we are working with the sine function, we will choose the default sine range.

[RANGE] [MENU] [D] [1]

The calculator automatically sets the X Range from $-\pi$ to π and the Y Range from -1.55 to 1.55.

```
X RANGE
Xmin=
    -6.283185307
Xmax=
     6.283185307
Xscl=
     1.570796327
```

Press the Graph mode key (top row) to draw the graph.

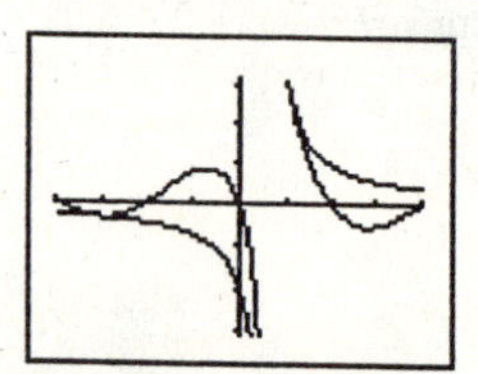

The calculator also has an Auto Range feature which will automatically set the Y range and draw the graph. This feature is very helpful if the function does not appear in the selected Range. To use the Auto Range press [2ndF] [AUTO].

.33 The Graph Menu

There are several options available for drawing the graphs and these can be used by pressing the [MENU] key when viewing the graph or equations. The first option allows you to quickly jump between the equation, fill and graph screens. Other options allow you to turn on a derivative trace (see section .34), draw the graphs without connecting the dots, and draw the graphs simultaneously.

The LOAD, SAVE, and DEL options are used to store up to 99 sets of equations. When you select Store, all of the 4 equations, the Range, the Menu settings and the Set Up of the calculator are all stored under one name.

.34 Tracing

To Trace along the curve, simple press the left or right arrow keys and Trace is automatically started. The Trace position is shown by the flashing cross, and its exact coordinates are given at the bottom of the screen. Press [CL] to clear the Trace mode. The up and down arrow keys move the cursor between Y1 and Y4. The cursor does <u>not</u> wrap from Y4 to Y1 (or vice versa) when the down arrow is pressed for the 4th time. This makes it easier to find which function you are tracing.

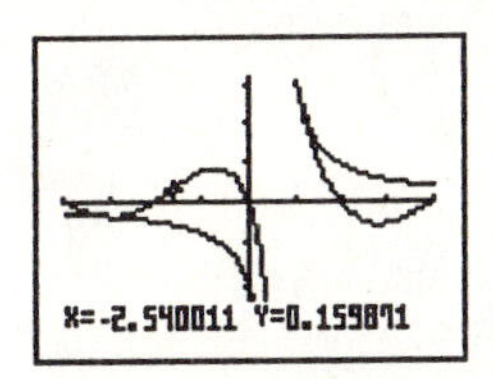

To quickly move to the right side of the screen press [2ndF] [>]. Use [2ndF] [<] to go back to the left side. You may notice that when the cursor moves off the screen, the graph will scroll in the X and Y directions to keep the cursor on the screen.

You can also get a trace value for the derivative of the function by pressing [MENU] [B] [1]. This function uses a numerical method to estimate the derivative at the X value.

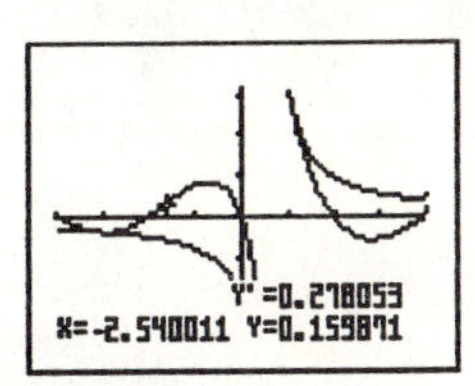

.35 Jumping

An easy way to zero in to points of interest is to use the Jump function. Pressing [2ndF] [JUMP] shows 5 Jump choices. The calculator will begin at the cursor position and work to the right looking for the selected point of interest on the graph. The cursor does not need to be in the vicinity of the point to have the calculator find it. The Trace function does not need to be active to use the Jump function. If there is no point found, the calculator will display the message 'NO SOLUTION IN RANGE".

Find the first intersection of the two equations:

[2ndF] [JUMP] [1]

The first intersection occurs at x=-4.712388.

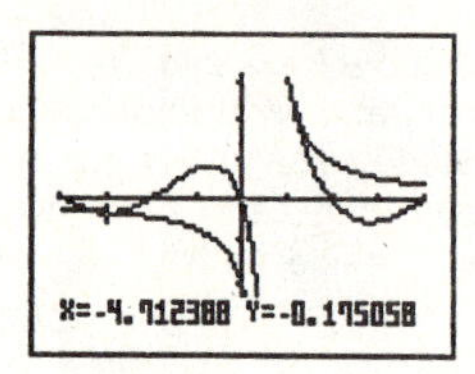

Find the first maximum of Y1:

[^] (this puts the cursor on Y1) [2ndF] [JUMP] [3]

The maximum occurs at X=-1.132267. If you are using the Y' Trace feature, you will notice that Y' is not exactly zero since numerical approximations were used to find the maximum and calculate Y'.

.36 Zooming

If you want to get a closer look at a part of the graph, you can use the Zoom feature. Press the [ZOOM] key and choose the Zoom option you want. Zoom Box allows you to draw a box on the graph which will become the new window. Zoom In and Zoom Out scale the graph by an amount specified in Zoom Factor. Zoom Auto sets the Y Range automatically. This is the same function as [2ndF] [AUTO].

Zoom in on the left part of the two curves:

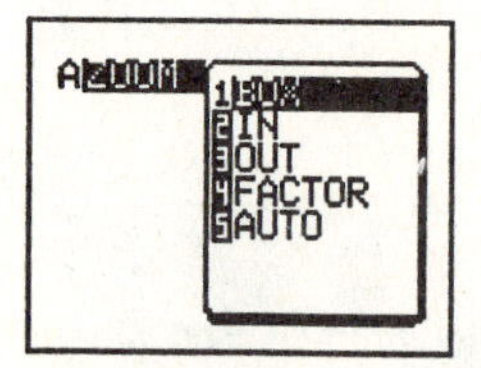

[ZOOM] [1] use the arrow keys to mark the first corner of the box [ENTER]. Use the arrow keys to draw the box. This will be your new viewing window.

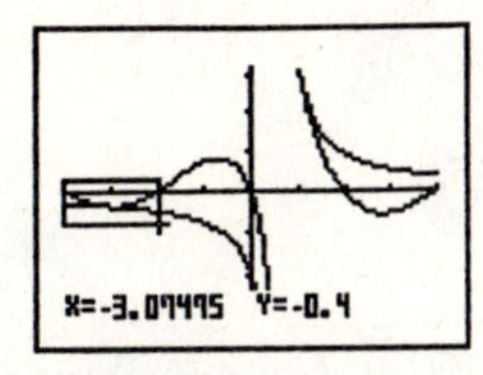

Press [ENTER] to redraw the graph.

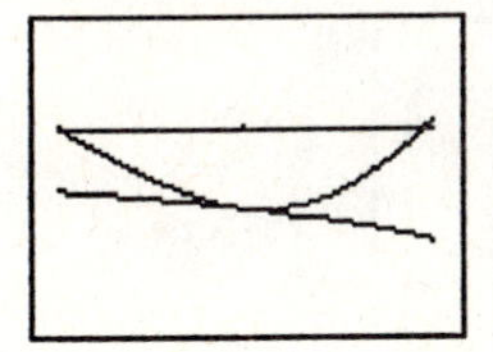

Matrix mode

Matrix mode lets you perform matrix operations and define matrix elements.

While in calculation mode, press [MENU] [A]. Enter matrix mode by pressing 3, or use the cursor keys and press [ENTER].

Using the Calculator

Defining a matrix

After entering matrix mode, press [MENU] [C] and the following menu appears:

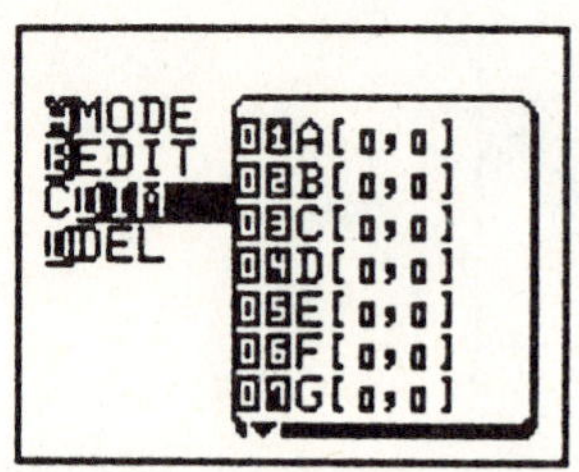

This option lets you define (and store) up to 26 different matrices. The matrices are stored even after the mode has been changed or the calculator has been turned off. To define the dimensions of a matrix, press 01 (or use the cursor keys to select a matrix and press [ENTER]). The following display appears:

Note: Redefining a matrix that already exists erases all of the information in that matrix. If you don't want to erase an existing matrix, press [QUIT] and select a different matrix.

Enter the number of rows and the number of columns. For example, to define a matrix with four rows and three columns, press 4 [ENTER] 3 [ENTER]

The first column of the matrix appears:

A[1,1]=
0.
A[2,1]=
0.
A[3,1]=
0.
A[4,1]=
0.

Use the keypad to enter the matrix elements and press [ENTER]. Each element of the matrix is labeled with its associated coordinate (matrix index [row number, column number]). Use the cursor keys to move around the matrix (through the row and column elements). Arrow indicators show you which directions you can move. [▼] and [▲] move through the rows. [◄] and [►] move through the columns. You can enter numbers, variables, and equations for each matrix element. (The final calculated value is stored.) Press [QUIT] to exit the matrix.

Example: Find the determinant of: $\begin{vmatrix} 1 & 3 \\ 2 & 4 \end{vmatrix}$

Press: [MENU] [C] [0] [1] 2 [ENTER] 2 [ENTER] (Defines a 2x2 matrix)

1 [ENTER] 2 [ENTER] 3 [ENTER] 4 [ENTER] [QUIT]

[MATH] [E] [6] [MAT] A [ENTER] ([2ndF] is not required)

Result: -2.

Editing a matrix

After entering matrix mode, press [MENU] [B] and the following menu appears:

To edit a matrix, enter its two digit number, or use the cursor keys to select a matrix and press [ENTER]. The first column of the matrix appears. Use the cursor keys to position the cursor on a matrix element, enter a new number, variable, or equation and press [ENTER].

Using the Calculator

Example: Change Matrix A $\begin{vmatrix} 1 & 3 \\ 2 & 4 \end{vmatrix}$ to $\begin{vmatrix} 11 & 3 \\ 2 & 4 \end{vmatrix}$

Press: [MENU] [B] [0] [1]

Press: 11 [ENTER] [QUIT]

Result:

```
A[1,1]=
            11.
A[2,1]=
             2.
```

Press: [▶]

Result:

```
A[1,2]=
             3.
A[2,2]=
             4.
```

Press: [QUIT]

Deleting a matrix

After entering matrix mode, press [MENU] [D] and the following menu appears:

To delete a matrix, enter its number, or select it with the cursor keys and press [ENTER].

Example: Delete matrix B

Press: [MENU] [D] [0] [2]

[ENTER] to delete the matrix, press [QUIT] to exit.

Result: If you pressed [ENTER], the matrix was deleted.

Matrix function keys

Except for [i] and [∠], all of the keyboard function keys are available in matrix mode.

[MAT] is the matrix identifier function. This function key tells the calculator that a matrix is being referenced. After pressing [MAT], you do not need to press [ALPHA] to select the matrix index. Use the matrix identifier to specify matrices used in calculations.

Example: Calculate Matrix A: $\begin{vmatrix} 11 & 3 \\ 2 & 4 \end{vmatrix}$ + Matrix B: $\begin{vmatrix} 1 & 16 \\ 31 & 41 \end{vmatrix}$

(First, create the above matrices using the steps described earlier.)

Press: [MAT] A [+] [MAT] B

Result:

```
ANS[1,1]=
        12.
ANS[2,1]=
        33.
```

Press: [▶]

Result:

```
ANS[1,2]=
        19.
ANS[2,2]=
        45.
```

Press: [QUIT]

Matrix 27 (ANS) is reserved for matrix operation results. Press [QUIT] to exit the ANS (or any other) matrix.

Two of the function keys ([x^{-1}] and [x^2]) perform operations that are very different from their operation in other modes. [x^{-1}] calculates the inverse of a square matrix. [x^2] calculates the square of a square matrix. Matrix A divided by Matrix B is the same as Matrix A x Matrix B^{-1}.

Texas Instruments TI–81

Operating the TI-81

The Keyboard

The keys on the TI-81 keyboard are divided into four zones: graphing keys, editing keys, advanced function keys, and scientific calculator keys.

Graphing Keys These keys are most frequently used to access the interactive graphing features of the TI–81.

Editing Keys These keys are most frequently used for editing expressions and values.

Advanced Function Keys These keys are used to access the advanced functions of the TI–81.

Scientific Calculator Keys These keys are used to access the capabilities of a standard scientific calculator.

This information is from the *TI–81Graphics Calculator Guidebook* published by Texas Instruments Incorporated in 1990 and is used with permission.

The Zones of the Keyboard

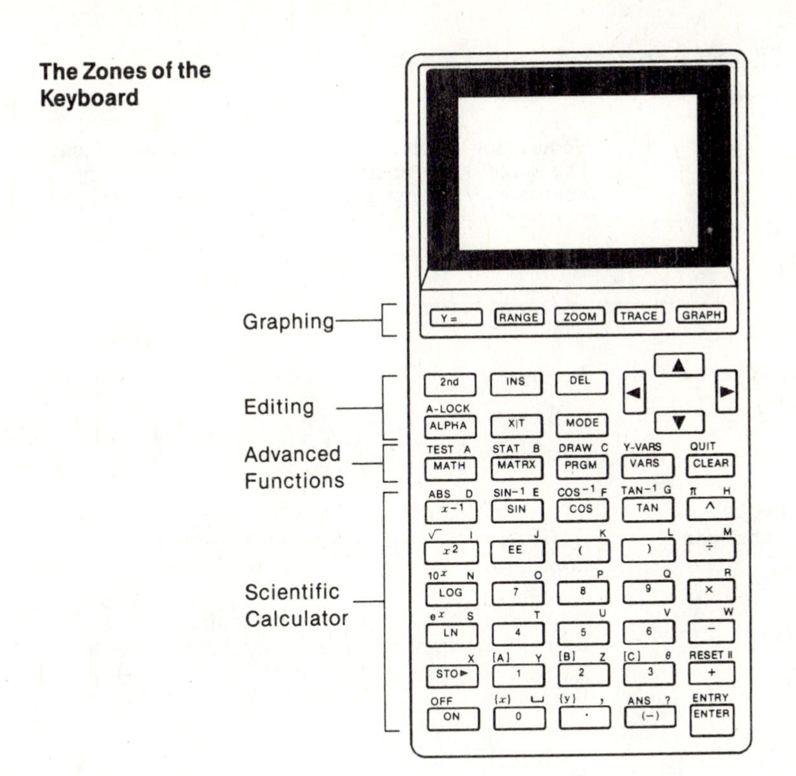

The Second and Alpha Keys

Some of the keys provide access to more than one function. These additional functions are printed above the keys and are accessed by first pressing the [2nd] or [ALPHA] key.

Key Labels

Second function —— √‾ I —— Alpha function

[x^2] —— Primary function

Second Functions

The operation printed to the left above a key is accessed by first pressing the [2nd] key and then pressing the appropriate key.

When you press the [2nd] key, the cursor changes to a highlighted up-arrow to indicate that the next keystroke is a second function. You can cancel second by pressing [2nd] again.

In this manual, second functions are shown in brackets and preceded by the [2nd] key symbol; for example, [2nd] [√‾].

Alphabetical Characters

The letter or symbol printed to the right above a key is accessed by first pressing the [ALPHA] key and then pressing the appropriate key.

When you press the [ALPHA] key, the cursor changes to a highlighted A to indicate that the next keystroke is an alphabetical character. You can cancel alpha by pressing [ALPHA] again.

In this manual, alphabetical characters are shown in brackets and preceded by the [ALPHA] key symbol; for example, [ALPHA] [I].

Alpha-Lock

You can press [2nd] [A-LOCK] to set alpha-lock, which makes each subsequent key press an alpha character. This is useful when entering display text in programs, for example, so that you will not need to press [ALPHA] before every letter. Cancel alpha-lock by pressing the [ALPHA] key.

Turning the TI-81 On and Off

To turn the TI-81 on, press the [ON] key. To turn the TI-81 off, press [2nd] [OFF]. After about five minutes without any activity, the APD™ Automatic Power Down feature turns the TI-81 off automatically.

Turning the TI-81 On

Press [ON] to turn the TI-81 on.

- If you turned the TI-81 off by pressing [2nd] [OFF], the display shows the Home screen with the cursor in the top left corner.
- If the APD™ feature turned the calculator off, the TI-81, including the display, cursor, and any error conditions, will be exactly as you left it.

Turning the TI-81 Off

Before turning the TI-81 off, be certain that you have saved any expressions or values that you want to recall later.

To turn the TI-81 off, press [2nd] [OFF].

- The display is cleared.
- Any error condition is cleared.
- Stored variables, programs, MODE settings, RANGE variables, contrast setting, Last Answer, Last Entry, and the most recent graph are retained in memory by the Constant Memory™ feature.

The APD™ Automatic Power Down Feature

To prolong the life of the batteries, the APD feature turns the TI-81 off automatically after about five minutes without any activity. When you press [ON], the TI-81 will be exactly as you left it.

- The display, cursor, and any error conditions are exactly as you left them.
- Stored variables, contrast setting, Last Answer, Last Entry, programs, MODE settings, RANGE variables, and the most recent graph are retained in memory.

Setting the Display Contrast

The brightness and contrast of the display depend on room lighting, battery freshness, viewing angle, and adjustment of the display contrast. The contrast setting is retained in memory when the TI-81 is turned off.

Adjusting the Display Contrast

You can adjust the display contrast to suit your viewing angle and lighting conditions at any time. As you change the contrast setting, the display contrast changes, and a number in the upper right corner between 0 (lightest) and 9 (darkest) indicates the current contrast setting.

To adjust the contrast:

1. Press the [2nd] key.
2. Use one of two keys:
   - To increase the contrast to the setting that you want, press and hold [▲].
   - To decrease the contrast to the setting that you want, press and hold [▼].

Caution: If you adjust the contrast setting to zero, the display may become completely blank. If this happens, press [2nd] and then press and hold [▲] until the display reappears.

When to Replace Batteries

When the batteries are low, the display begins to dim (especially during calculations), and you must adjust the contrast to a higher setting. If you find it necessary to set the contrast to a setting of 8 or 9, you should replace the batteries soon.

The Display

The TI-81 displays both text and graphs. When text is displayed, the screen can display up to eight lines of 16 characters per line. When all eight lines of the screen are filled, text "scrolls" off the top of the screen.

The Home Screen When you turn the TI-81 on, the Home screen is displayed. The Home screen is the primary screen of the TI-81. On it you enter expressions and instructions and see the results.

```
17*3+ln 3          — Expression
     52.09861229   — Result
```

Display Cursors The TI-81 has several special cursors. In most cases, the appearance of the cursor indicates what will happen when you press the next key.

The cursors that you see on the Home screen are described here. Other special cursors are described in the appropriate chapters.

Cursor	Appearance	Meaning
Entry cursor	Solid blinking rectangle	The next keystroke is entered at the cursor, overwriting any character
Insert cursor	Blinking underline	The next keystroke is inserted at the cursor
2nd cursor	Highlighted blinking ↑	The next keystroke is a second function
ALPHA cursor	Highlighted blinking **A**	The next keystroke is an alpha character

Busy Indicator When the TI-81 is calculating or graphing, a box in the upper right of the screen is highlighted.

Returning to the Home Screen You can return to the Home screen from any other screen by pressing [2nd] [QUIT].

Menu Screens

You can access functions and operations that are not on the keyboard through menus. Menu screens temporarily replace the screen where you are working. After you select an item from a menu, the screen where you are working is displayed again.

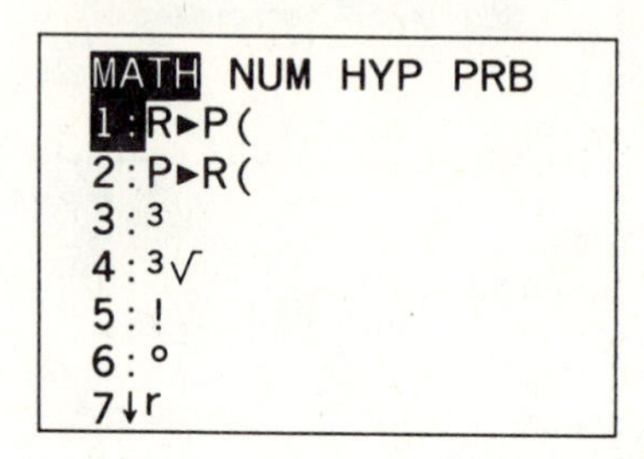

Graph Screens

The graph screen displays a graph of selected functions and the cursor coordinate values.

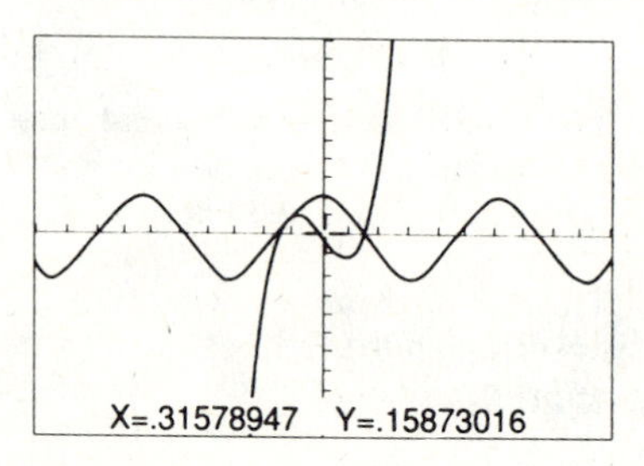

Editing Screens

There are several editing screens that are used to enter or edit expressions or values.

- Statistical data
- Matrix values
- Functions in the Y= list for graphing
- RANGE variables for the viewing rectangle
- Zoom factors for exploring graphs
- Programs

The Equation Operating System

The Equation Operating System (EOS) allows numbers and operations to be entered into the TI-81 in a simple, straightforward sequence. EOS evaluates expressions according to the standard priorities of mathematical operations and uses parentheses for grouping.

Order of Evaluation

EOS evaluates operations in an expression in the following order:

- Polar/rectangular conversions, numerical derivative, round, and row operations.
- Math operations and functions that are entered after the argument, such as x^2, x^{-1}, x^3, x!, °, ʳ, and transpose.
- Universal powers, such as y^x.
- Implied multiplication where the second argument is a number, a variable, or a matrix, such as 2π, 4B, 3[C], or sin (A+B)4.
- Math and trig functions that are entered before the argument, such as sine, cosine, tangent, and their inverses, log and antilog, natural log and antilog, absolute value, square root, negation, greatest integer, integer part, fractional part, cube root, determinant, and hyperbolic functions.
- Implied multiplication other than above, such as 3log 4 or sin 4(A+B).
- Permutations and combinations (nPr and nCr).
- Multiplication and division.
- Addition and subtraction.
- Relational operators, such as $>$ or $\leq$.

Within a priority group, EOS evaluates operations from left to right. Calculations inside a pair of parentheses are evaluated first.

When you press [ENTER], regardless of the cursor location, the expression is completed; the cursor does not need to be at the end of the expression.

Implied Multiplication

When a number precedes a variable, matrix, math or trig function, left parenthesis, or π, it is not necessary to enter the multiplication sign. The TI-81 understands 2π, 4sin 45, 5(1+2), (2∗5)(7−4), or AB, for example, as implied multiplication. See the previous page for how the TI-81 evaluates implied multiplication.

Using Parentheses

All calculations inside a pair of parentheses are done first. Then the result of that calculation is used to continue the evaluation.

For example, in the expression (1+2)4, the TI-81 first evaluates the portion of the expression inside the parentheses, 1+2, and then multiplies the result, 3, by 4.

You may omit any right (close) parenthesis at the end of an expression. All "open" parenthetical elements are closed automatically at the end of the expression when you press [ENTER]. For clarity, however, all close parentheses are shown in this manual.

Entering a Negative Number

To enter a negative number:

1. Press [(−)].

2. Enter the number.

For example, press [(−)] 1.7 to enter −1.7.

Note: The [−] key is used for subtraction. The TI-81 displays the error message **ERROR 06 SYNTAX** for either of the following cases:

- If you press [−] for a negative number, as in 9 [×] [−] 7 [ENTER].
- If you press [(−)] to indicate subtraction, as in 9 [(−)] 7 [ENTER].

Entering Expressions for Evaluation

An expression is a complete sequence of numbers, operations, variables, functions, and their arguments that can be evaluated to a single result. On the TI-81, an expression is entered in the same order that it normally is written.

Entering an Expression

You enter numbers, variable names symbols, functions, and operations from the keyboard and from menus to create an expression. An expression is completed when you press [ENTER], regardless of the cursor location. The entire expression is evaluated according to EOS, and the result is displayed.

Notice that most of the functions and operations on the keyboard and menus are symbols with several characters in them. You must enter the symbol from the keyboard or menu, not spell it out. For example, to calculate the log of 45, you must press [LOG] 45; you cannot type in the letters LOG. (If you type LOG, the TI-81 interprets the entry as implied multiplication of the variables L, O, and G.)

Example of Entering an Expression

Calculate $3.76 \div (-7.9 + \sqrt{5}) + 2 \log 45$.

Procedure	Keystrokes	Display
Begin expression	3.76 [÷]	3.76/
Begin parentheses	[(]	3.76/(
Enter negative 7.9	[(-)] 7.9	3.76/(-7.9
Add square root of 5	[+] [2nd] [√] 5	3.76/(-7.9+√5
Complete parentheses	[)]	3.76/(-7.9+√5)
Add 2 log 45	[+] 2 [LOG] 45	3.76/(-7.9+√5)+2 log 45
Evaluate expression	[ENTER]	3.76/(-7.9+√5)+2 log 45 2.642575252

Continuing an Expression

You can recall the answer to a calculation as the first entry in the next expression without reentering the value. Simply begin by pressing the operation key. The TI–81 inserts the variable **Ans** , which contains the last answer, into the expression.

Example of Continuing an Expression

Square the result of the example on the previous page.

Procedure	Keystrokes	Display
Square prior result	[x^2]	Ans^2
Evaluate expression	[ENTER]	Ans^2 6.983203964

Notes about Entering Expressions

Notice the following about entering expressions:

- Sometimes the displayed symbol is not the same as the key symbol, as in x^2, e^x, and ÷.
- The TI–81 interprets angles in trig functions based on the MODE setting (degrees or radians).
- If an expression is longer than 16 characters, it "wraps" around to the beginning of the next line.
- The result is displayed on the right side of the next line. The result can be up to ten digits with a two-digit exponent. The MODE settings determine the notation format and number of decimal places displayed.

Editing Expressions

Expressions can be edited. The cursor-movement keys move the cursor within and between lines. Normal entry types over the character or symbol where the cursor is. The [INS] and [DEL] keys insert or delete characters or symbols.

The Cursor-Movement Keys

The arrow keys in the upper right of the keyboard control the movement of the cursor.

The [◄] and [►] keys move the cursor within an expression. The cursor stops when it reaches the beginning or end of the expression.

The [▼] and [▲] keys move the cursor between lines.

When you press and hold one of the cursor-movement keys, the cursor movement repeats until you release the key.

The Edit Keys

Key	Meaning
[INS]	Inserts characters or symbols at the blinking underline cursor
[DEL]	Deletes the character or symbol at the blinking cursor
[CLEAR]	Clears (blanks) the entire expression; on the Home screen, it clears the screen
[ENTER]	Completes the expression

When you press [2nd] or [ALPHA] during an insert, the underline cursor changes to an underlined ↑ or **A** cursor.

Inserting in Expressions

To insert a character or symbol in an expression:

1. Use the cursor-movement keys to position the cursor on the character or symbol in front of which you want to insert.

2. Press [INS].

   The cursor changes to a blinking underline.

3. Enter the characters or symbols you want to insert.

4. End the insert in one of the following ways:

   - Press [INS] again.
   - Press a cursor-movement key.

Deleting from Expressions

To delete a character or symbol from an expression:

1. Use the cursor-movement keys to position the cursor on the character or symbol you want to delete.

2. Press [DEL].

   The character or symbol is deleted. All the characters of a symbol that is represented by a group of characters (such as **log** or **sin**) are deleted together.

Setting Modes

Modes determine how numbers and graphs are displayed and calculated. MODE settings are retained by the Constant Memory™ feature when the TI-81 is off.

Checking MODE Settings

Press the [MODE] key to display the MODE settings. The current settings are highlighted. The various MODE settings are described on the following pages.

Setting	Meaning
Norm Sci Eng	Type of notation for display
Float 0123456789	Number of decimal places
Rad Deg	Type of angle measure
Function Param	Function or parametric graphing
Connected Dot	Whether to connect plotted points
Sequence Simul	How to plot selected functions
Grid Off Grid On	Whether to display a graph grid
Rect Polar	Type of graph coordinate display

Changing MODE Settings

To change any of the settings:

1. Use [▼] or [▲] to move the cursor to the row of the setting that you want to change. The setting that the cursor is on blinks.

2. Use [▶] or [◀] to move the cursor to the setting that you want.

3. Press [ENTER] to select the blinking setting.

Leaving the MODE Screen

When the MODE settings are as you want them, leave the MODE screen in one of the following ways:

- Select another screen by pressing the appropriate key, such as [Y=] or [GRAPH].

- Press [2nd] [QUIT] to return to the Home screen.

Normal, Scientific, or Engineering Notation Display Format

Notation formats affect only how a numeric result is displayed. You can enter a number in any format.

Normal display format is the way in which we usually express numbers, with digits to the left and right of the decimal, as in 12345.67.

Scientific notation expresses numbers in two parts. The significant digits are displayed with one digit to the left of the decimal. The appropriate power of 10 is displayed to the right of E, as in 1.234567E4.

Engineering notation is similar to scientific notation. However, the number may have one, two, or three digits before the decimal, and the power-of-10 exponent is a multiple of three, as in 12.34567E3.

Note: If you select normal display format, but the result cannot be displayed in 10 digits or the absolute value is less than .001, the TI–81 switches to a scientific format for that result only.

Floating or Fixed Decimal Display Setting

Decimal settings affect only how a result is displayed. They apply to all three notation display formats.

Floating decimal setting displays up to 10 digits, plus the sign and decimal.

Fixed decimal setting displays the selected number of digits to the right of the decimal. Place the cursor on the number of decimal digits you want and then press [ENTER].

Radians or Degrees Angle Setting

Radian setting means that angle arguments in trig functions or polar/rectangular conversions are interpreted as radians. Results display in radians.

Degree setting means that angle arguments in trig functions or polar/rectangular conversions are interpreted as degrees. Results display in degrees.

Defining a Matrix

To define a matrix, select one of the three matrices, [A], [B], or [C], and define the dimensions of that matrix. On the top line of the MATRX edit screen, a small graphic box indicates the size of the matrix and the element where the cursor is located.

Selecting a Matrix

To define a matrix, you first must select the matrix you want to define.

1. Press [MATRX] [▶] to select the MATRX EDIT menu.
2. Press the number of the matrix you want to create (either [A], [B], or [C]). The MATRX edit screen appears.

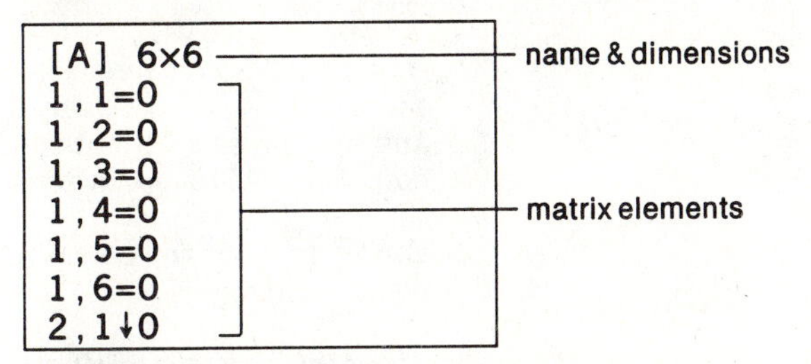

The blinking cursor is on the row dimension. A ↓ is displayed in place of the = on the last line if there are more than seven elements in the matrix.

Accepting or Changing Matrix Dimensions

The dimensions of the matrix (row × column) are displayed on the top line. You must accept or change the dimensions each time you enter or edit a matrix.

1. Accept or change the number of rows.
   - To accept the number, press [ENTER].
   - To change the number, enter the number of rows (up to six), and then press [ENTER].

   The cursor moves to the number of columns.

2. Accept or change the number of columns.
   - To accept the number, press [ENTER].
   - To change the number, enter the number of columns (up to six), and then press [ENTER].

   The cursor moves to the first matrix element (1,1).

Entering and Editing Matrix Elements

After the dimensions of the matrix are set, values can be entered into the matrix elements on the MATRX edit screen. On the top line of the screen, a small graphic box indicates the size of the matrix and the element where the cursor is located.

Entering Matrix Elements

To enter values in a matrix:

1. Enter the value you want and then press [ENTER]. The cursor moves to the next element.

2. Continue entering values.

Editing a Matrix

To edit a matrix:

1. If the MATRX edit screen is not currently displayed, press [MATRX] [▶] to display the MATRX EDIT menu, and then press the number of the matrix you want to edit. The MATRX edit screen appears.

2. Use [▼] to move the cursor to the matrix element you want to change.

3. Enter the new value using one of these methods:

   - Enter a new value. The original value is cleared automatically when you begin typing.

   - Use [▶] or [◀] to position the cursor over the digit you want to change. Then type over it or use [DEL] to delete it.

4. When you have changed the matrix element value, press [ENTER]. The cursor moves to the next value.

Leaving the MATRX Edit Screen

When you finish entering and editing matrix values:

- Select another screen by pressing the appropriate key, such as [MATRX].

- Press [2nd] [QUIT] to return to the Home screen.

Clearing a Matrix

To clear all the elements of a matrix, store 0 to that matrix. For example, press 0 [STO▶] [2nd] [[A]] to set all the elements in matrix [A] to zero.

Notes about Using Matrices

Matrices can be displayed, copied, and used in the same way as a variable.

Displaying a Stored Matrix

You can display a stored matrix in one of two ways:

- On the Home screen, press [2nd] [[A]], for example, and then press [ENTER]. The matrix is displayed in matrix format:

```
[A]
[ 1 0 1 0]
[ 1 2 0 0]
[ 3 1 1 1]
```

  If all of the matrix does not fit in the display, as indicated by dots on the right of the screen, use [▶] and [◀] to display the rest of the matrix.

- Press [MATRX] [▶] to select the MATRX EDIT menu, and then press the number of the matrix you want. The MATRX edit screen appears (see page 6–3). Use [▼] to view the elements.

Copying One Matrix to Another

Copy one matrix to another by storing the matrix to another matrix location. For example, press [2nd] [[A]] [STO▶] [2nd] [[B]] to copy [A] to [B]. The dimensions of [B] will be the dimensions of [A].

Using a Matrix in an Expression

To enter the name of a matrix into an expression, press [2nd] [[A]], [2nd] [[B]], or [2nd] [[C]]. The name of the matrix is copied to the current cursor location in the expression you are editing.

Results of Matrix Calculations

If the result of evaluating an expression is a matrix, it is stored in the variable **Ans**. To save the resulting matrix, store **Ans** to a matrix. For example, press [2nd] [ANS] [STO▶] [2nd] [[A]] to store the resulting matrix to matrix [A].

Texas Instruments TI–85

Operating the TI-85

Turning the TI-85 On and Off

To turn the TI-85 on, press the [ON] key. To turn it off, press and release [2nd] and then press [OFF]. After about five minutes without any activity, the APD™ Automatic Power Down feature turns the TI-85 off automatically.

Turning the Calculator On

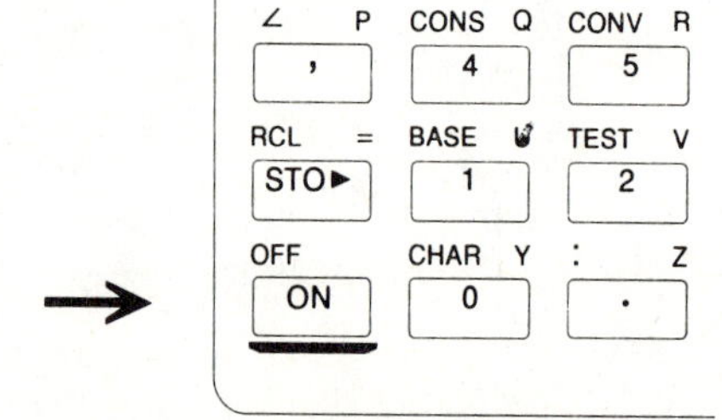

Press [ON] to turn the TI-85 on.

- If you pressed [2nd] [OFF] to turn the calculator off, the display shows the Home screen as it was when you last used it.
- If the APD feature turned the calculator off, the TI-85, including the display, cursor, and any error conditions, will be exactly as you left it.

Turning the Calculator Off

Press and release [2nd] and then press [OFF] to turn the TI-85 off.

- Any error condition is cleared.
- All settings and memory contents are retained in memory by the Constant Memory™ feature.

The APD™ Automatic Power Down Feature

To prolong the life of the batteries, the APD feature turns the TI-85 off automatically after about five minutes without any activity. When you press [ON], the TI-85 will be exactly as you left it.

- The display, cursor, and any error conditions are exactly as you left them.

This information is from the *TI–85 Advanced Scientific Calculator Guidebook* published by Texas Instruments Incorporated in 1992 and is used with permission.

- All settings and memory contents are retained in memory by the Constant Memory feature.

Batteries

The TI-85 uses four AAA alkaline batteries and has a user-replaceable back-up lithium battery. You can change the batteries without losing any information in memory.

Setting the Display Contrast

The brightness and contrast of the display depend on room lighting, battery freshness, viewing angle, and adjustment of the display contrast. The contrast setting is retained in memory when the TI-85 is turned off.

Adjusting the Display Contrast

You can adjust the display contrast to suit your viewing angle and lighting conditions at any time. As you change the contrast setting, the display contrast changes, and a number in the upper right corner indicates the current contrast setting between 0 (lightest) and 9 (darkest).

To adjust the contrast:

1. Press and release the [2nd] key.
2. Use one of two keys:
   - To increase the contrast, press and hold [▲].
   - To decrease the contrast, press and hold [▼].

Note: If you adjust the contrast setting to zero, the display may become completely blank. If this happens, press and release [2nd] and then press and hold [▲] until the display reappears.

When to Replace Batteries

When the batteries are low, the display begins to dim (especially during calculations), and you must adjust the contrast to a higher setting. If you find it necessary to set the contrast to a setting of 8 or 9, you should replace the four AAA batteries soon.

The 2nd and ALPHA Keys

Most keys on the TI-85 access more than one operation. The additional operations are printed above the keys. To access them, press [2nd] or [ALPHA] before you press the key.

Key Labels

2nd operation — √ K — Alpha operation

x^2 — Primary operation

2nd Operations

To access a 2nd operation, first press and release [2nd] and then press the appropriate key.

When you press [2nd], the cursor changes to ↑ to indicate that the next keystroke is a 2nd operation.

To cancel 2nd, press [2nd] again.

In this guidebook, 2nd operations are shown in brackets and preceded by [2nd]; for example, [2nd] [√].

ALPHA Characters

To access the letter or character printed to the right above a key, first press [ALPHA] or [2nd] [alpha] and then press the appropriate key.

- To make the next keystroke an uppercase alphabetic character, press [ALPHA]. The cursor changes to **A**. To cancel ALPHA, press [ALPHA] until the normal cursor appears.
- To make the next keystroke a lowercase alphabetic character, press and release [2nd] and then press [alpha]. The cursor changes to **a**. To cancel alpha, press [ALPHA] until the normal cursor appears.

Alpha-Lock

ALPHA-lock (uppercase) and alpha-lock (lowercase) make each subsequent keystroke an alphabetic character. You do not need to press [ALPHA] or [2nd] [alpha] before every character to enter display text or the names of variables, functions, or instructions.

Action	Keystrokes
Set uppercase ALPHA-lock	[ALPHA] [ALPHA]
Set lowercase alpha-lock	[2nd] [alpha] [ALPHA] or [2nd] [alpha] [2nd] [alpha] or [ALPHA] [2nd] [alpha]
Cancel ALPHA-lock	[ALPHA]
Cancel alpha-lock	[2nd] [alpha] or [ALPHA] [ALPHA]
Change from uppercase ALPHA-lock to lowercase alpha-lock	[2nd] [alpha]
Change from lowercase alpha-lock to uppercase ALPHA-lock	[ALPHA]

Note: [STO▶] and name prompts automatically set the keyboard in ALPHA-lock. [2nd] does not take the keyboard out of ALPHA-lock or alpha-lock.

The Display

The TI-85 displays text, graphs, and menus.

The Home Screen

The Home screen is the primary screen of the TI–85, where you enter expressions to be evaluated and see the results.

```
17*3+ln 3
       52.0986122887
```

Expression
Result

If text is displayed, the screen can have up to eight lines of 21 characters per line. If all text lines of the display are filled, text "scrolls" off the top of the display.

The MODE settings control the way expressions are interpreted and results are displayed.

Displaying Expressions

On the Home screen and in the program editor, if an expression is longer than one line, it wraps to the beginning of the next line.

Displaying Results

When an expression is evaluated on the Home screen, the result is displayed on the right side of the next line. If a result is too long to display in its entirety, ellipsis marks (...) are shown at the left or right. Use [►] and [◄] to scroll the result. If the result is a matrix with more rows than the screen can display, use [▲] and [▼] to scroll the result vertically. For example:

Expression
Result

Returning to the Home Screen

To return to the Home screen from any other screen, press [2nd] [QUIT].

Display Cursors

In most cases, the appearance of the cursor indicates what will happen when you press the next key.

Cursor	Appearance	Meaning
Entry cursor	Solid blinking rectangle	The next keystroke is entered at the cursor; it types over any character.
INS (insert) cursor	Blinking underline	The next keystroke is inserted at the cursor.
2nd cursor	Blinking ↑	The next keystroke is a 2nd operation.
ALPHA cursor	Blinking **A**	The next keystroke is an uppercase alphabetic character.
alpha cursor	Blinking **a**	The next keystroke is a lowercase alphabetic character.
"full" cursor	Checkerboard rectangle	You have entered the maximum characters in a name, or memory is full.

If you press [ALPHA], [2nd] [alpha], or [2nd] during an insertion, the underline cursor changes to an underlined **A**, **a**, or ↑ cursor.

Busy Indicator

When the TI-85 is calculating or graphing, a moving vertical bar shows in the upper right of the display as a busy indicator. (When you pause a graph or a program, the busy indicator is a dotted bar.)

The Equation Operating System

With the TI-85's Equation Operating System (EOS™), you enter numbers and functions in a simple, straightforward sequence. EOS evaluates expressions according to the standard priorities of mathematical functions and uses parentheses for grouping.

Order of Evaluation

A function returns a value. EOS evaluates functions in an expression in this order:

- Functions that are entered after the argument, such as **x²**, **x⁻¹**, **!**, **°**, **ʳ**, **%**, **ᵀ**, and conversions.
- Powers and roots, such as **2^5** or **5ˣ√32**.
- Implied multiplication where the second argument is a number, variable name, constant, list, matrix, or vector or begins with an open parenthesis, such as **4A**, **A B**, **(A+B)4**, or **4(A+B)**.
- Single-argument functions that precede the argument, such as negation, √, **sin**, or **ln**.
- Implied multiplication where the second argument is a multiargument function or a single-argument function that precedes the argument, such as **2 gcd(144,64)** or **A sin 2**.
- Permutations (**nPr**) and combinations (**nCr**).
- Multiplication and division.
- Addition and subtraction. An **=** in an expression, rather than an equation, is evaluated as **−(**. For example, **A+B=C+1** is evaluated as **A+B−(C+1)**.
- Relational functions, such as > or ≤.
- Boolean operator **and**.
- Boolean operators **or** and **xor**.

Within a priority group, EOS evaluates functions from left to right. However, two or more single-argument functions that precede the same argument are evaluated from right to left. For example, **sin fPart ln 8** is evaluated as **sin(fPart(ln 8))**.

Calculations within a pair of parentheses are evaluated first. Multiargument functions, such as **gcd(144,64)** or **der1(sin ANG,ANG,π)**, are evaluated as they are encountered.

Implied Multiplication

The TI–85 recognizes implied multiplication. For example, it understands **2π**, **4 sin 45**, **5(1 + 2)**, and **(2∗5)7** as implied multiplication. Except between two numbers, a space indicates implied multiplication, as in **A B** or **B 3**.

Variable names can be more than one character; the TI–85 recognizes **AB** and **b2** as variable names. Variable names cannot start with a number; **3AB** and **3b2** are interpreted as implied multiplication (**3∗AB** and **3∗b2**).

Parentheses

All calculations inside a pair of parentheses are completed first. For example, in the expression **4(1 + 2)**, EOS first evaluates the portion of the expression inside the parentheses, 1+2, and then multiplies the result, 3, by 4.

You can omit any right (close) parenthesis at the end of an expression. All "open" parenthetical elements are closed automatically at the end of an expression and preceding the → (store) or display conversion instructions.

Note: If the name of a list, matrix, or vector is followed by an open parenthesis, it does not indicate implied multiplication. It is used to access specific elements in the list, matrix, or vector.

Negation

To enter a negative number, use the negation function. Press [(-)] and then enter the number. On the TI–85, negation is in the fourth group in the EOS hierarchy. Functions in the first group, such as squaring, are evaluated before negation. For example, the result of **-X²** is a negative number; the result of **-9²** is **-81**. Use parentheses to square a negative number: **(-9)²**.

Note: Use the [-] key for subtraction and the [(-)] key for negation. If you press [-] to enter a negative number, as in **9** [×] [-] **7**, it is an error. If you press **9** [(-)] **7** or [ALPHA] **A** [(-)] [ALPHA] **B**, it is interpreted as implied multiplication (**9∗-7** or **A∗-B**).

Entering and Editing

The arrow keys in the upper right of the keyboard control the movement of the cursor. In normal entry, a keystroke types over the character or characters at the position of the cursor. The [DEL] and [2nd] [INS] keys delete or insert characters.

The Cursor-Movement Keys

[◄] and [►] move the cursor within an expression. The cursor stops when it reaches the beginning or end of the expression, except in the program editor.

[2nd] [◄] or [2nd] [►] moves the cursor to the beginning or end of the expression.

[▼] and [▲] move the cursor between lines in the current expression on the Home screen. [▲] on the top line of an expression on the Home screen moves the cursor to the beginning of the expression. [▼] on the bottom line moves the cursor to the end.

If you press and hold a cursor-movement key, the cursor movement repeats until you release the key.

The Edit Keys

Key	Action
[2nd] [INS]	Inserts characters at the underline cursor.
[DEL]	Deletes the character at the cursor.
[ENTER]	Executes the expression or instruction.
[CLEAR]	• On a line with text on the Home screen, clears (blanks) that line. • In an editor, clears (blanks) the expression or value where the cursor is located; it does not store a zero. • On a blank line on the Home screen, clears everything on the Home screen.

To end insertion, press [2nd] [INS], a cursor-movement key, [DEL], or (except in the program editor) [ENTER].

You can press and hold [DEL] to delete a long sequence of characters.

Entering a Name You can enter the names of functions, instructions, variables, and constants in one of several ways:

- Type the characters of the name.
- Press the key or select from a menu to copy the name to the cursor location.
- Select the name from the CATALOG.

If you type the name, you must enter each character, including a space (the alpha character above [(-)]) preceding the name and the space or open parenthesis, if required, after the name. If you select the name from the keyboard or a menu, all required characters are copied.

The TI-85 ignores uppercase and lowercase when it interprets names of functions and instructions (but not the names of variables and constants). For example, to calculate a log, you can press [LOG] , type the letters **l o g** (followed by a space), or type the letters **L O G** (followed by a space).

Character Entry The TI-85 treats an expression as individual characters, regardless of whether a name was entered by typing each character or by copying the name from a key, menu, or selection screen. Names copied from a key, menu, or selection screen are copied as if the individual letters were typed. You can type over any character in the name. For example, if you press [SIN] , the characters **sin** followed by a space are displayed. If you then press [◄] [◄] [ALPHA] [ALPHA] **G N**, the function is changed to **siGN**.

Expressions and Instructions

On the TI-85, you can enter expressions, which return a value, in most places where a value is required. You enter instructions, which initiate an action, on the Home screen or in the program editor

Expressions

An expression is a complete sequence of numbers, variables, functions, and their arguments that evaluate to a single result. On the TI–85, you enter an expression in the same order that it normally is written. For example, π∗**radius²** is an expression.

Expressions can be used as commands on the Home screen to calculate a result. Expressions may be used in instructions to enter a value. In editors, expressions may be used to enter a value.

Instructions

An instruction is a command that initiates an action. For example, **ClDrw** is an instruction that clears any drawn elements from a graph. Instructions cannot be used in expressions.

Entering an Expression

To create an expression, you enter numbers, variables, and functions from the keyboard and from display menus. An expression is completed when you press [ENTER], regardless of the cursor location. The entire expression is evaluated according to EOS, and the result is displayed.

Example of Entering an Expression

Calculate 3.76 ÷ (-7.9 + √5) + 2 log 45.

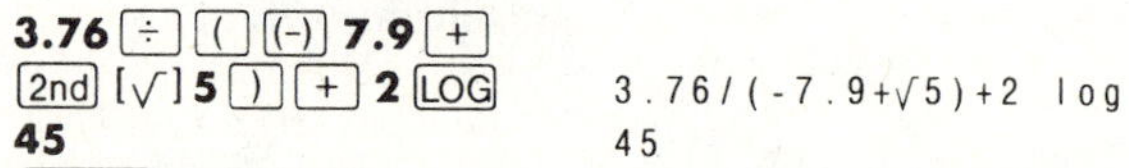

3.76 [÷] [(] [(-)] **7.9** [+]
[2nd] [√] **5** [)] [+] **2** [LOG] 3.76/(-7.9+√5)+2 log
45 45
[ENTER] 2.64257525233

Entering More than One Command on a Line

To enter more than one instruction or expression on a line, separate them with a colon (**:**). For example, **5→A:2→B:A/B** [ENTER] displays **2.5**. All the commands are stored together in Last Entry.

Interrupting a Calculation

While the busy indicator is displayed, indicating that a calculation or a graph is in progress, you can press [ON] to stop the calculation. (There may be a delay.) Except in graphing, the break ERROR screen is shown.

- To go to where the interrupt occurred, select 〈GOTO〉.
- To return to the Home screen, select 〈QUIT〉.

Last Answer

When an expression is evaluated successfully from the Home screen or from a program, the TI-85 stores the result to a special variable, Ans (Last Answer). When you turn the TI-85 off, the value in Ans is retained in memory.

Using Last Answer in an Expression

You can use the variable **Ans** in most places where its data type is valid. Press [2nd] [ANS] and the variable name **Ans** is copied to the cursor location. When the expression is evaluated, the TI-85 uses the value of **Ans** in the calculation.

Calculate the volume of a cube 1.5 feet on each side, and then calculate the volume in cubic inches.

Keys	Display
1.5 [^] **3**	1.5^3
[ENTER]	3.375
12 [^] **3** [2nd] [ANS]	12^3 Ans
[ENTER]	5832

Continuing an Expression

You can use the value **Ans** as the first entry in the next expression without entering the value again. On the blank line on the Home screen, enter the function; the TI-85 "types" the variable name **Ans** followed by the function.

Calculate the area of a circle of radius 5 inches. Then calculate the volume of a cylinder of height 3 inches and radius 5 inches.

Keys	Display
[2nd] [π] **5** [x^2]	π5²
[ENTER]	78.5398163397
[×] **3**	Ans*3
[ENTER]	235.619449019

Storing Results

To store a result, store **Ans** to a variable before you evaluate another expression.

Keys	Display
[STO►] **VOLUME**	Ans→VOLUME
[ENTER]	235.619449019

Last Entry

When you press [ENTER] on the Home screen to evaluate an expression or execute an instruction, the expression or instruction is stored in a special storage area called Last Entry, which you can recall. When you turn the TI-85 off, Last Entry is retained in memory.

Using Last Entry

To recall Last Entry and edit it, press [2nd] [ENTRY]. The cursor is positioned at the end of the entry. Because the TI-85 updates the Last Entry storage area only when [ENTER] is pressed, you can recall the previous entry even if you have begun entering the next expression. However, when you recall Last Entry, it replaces what you have typed.

Keys	Display
5 [+] **7**	5+7
[ENTER]	12
[2nd] [ENTRY]	5+7

Entries Containing More than One Command

If the previous entry contained more than one command separated with a colon (page 1-12), all the commands are recalled. You can recall all commands, edit any command, and then execute all commands.

Using the equation $A = \pi r^2$, find by trial and error the radius of a circle that covers 200 square inches. Use 8 as your first guess.

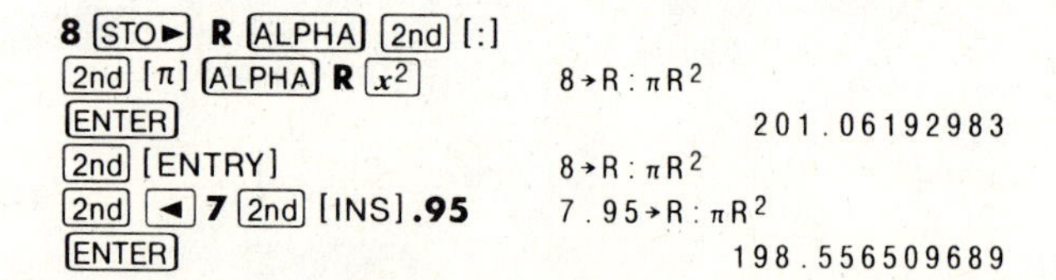

Continue until the result is as accurate as you want.

Reexecuting the Previous Entry

Press [ENTER] on a blank line on the Home screen to execute Last Entry; the entry does not display again.

Keys	Display
0 [STO►] **N**	0→N
[ENTER]	0
[ALPHA] **N** [+] **1** [STO►] **N**	
[2nd] [:] **N** [ALPHA] [x^2] [−] **1**	N+1→N:N²−1
[ENTER]	0
[ENTER]	3
[ENTER]	8

Example: Convergence of a Series

Show that when A<1, the series A^N converges to A/(1 – A) as N gets large. You can use the TI-85 functions sum and seq to calculate a series.

Procedure

Calculate the series $\mathbf{A^N}$ for **A** = 1/2 at **N** = 1, 5, and 100. **sum** returns the sum of all elements in a list. **seq** generates a list; the form for **seq** is:

seq(*expression***,***variablename***,***begin***,***end***,***increment***)**

Enter all expressions and instructions on the same command line so that you can recall, edit, and execute them. Store **1** to the variable **NTH** (for the *nth* element) and **1/2** to the variable **A**.

Remember that function names are not case-sensitive, but variable names are. The keyboard remains in ALPHA-lock after [STO►], even when you press [2nd].

Keystrokes	Display
1 [STO►] **NTH** [2nd] [:] [ALPHA]	
1 [÷] **2** [STO►] **A** [2nd] [:]	
SEQ [ALPHA] [(] [ALPHA] **A**	
[^] [ALPHA] **N** [,] [ALPHA] **N**	1→NTH:1/2→A:SEQ(A^N,N
[,] **1** [,] [ALPHA] [ALPHA]	
NTH [ALPHA] [,] **1** [)] [STO►]	
LIST [2nd] [:] **SUM** [␣] **LI**	,1,NTH,1)→LIST:SUM LI
ST	ST
[ENTER]	.5

Recall Last Entry. Change **NTH** to **5** and evaluate. Repeat for **NTH** = **100**.

Keystrokes	Display
[2nd] [ENTRY]	1→NTH:1/2→A:seq(A^N,N ,1,NTH,1)→LIST:sum LI ST
[2nd] [◄] **5**	5→NTH:1/2→A:seq(A^N,N ,1,NTH,1)→LIST:sum LI ST
[ENTER]	.96875
[2nd] [ENTRY] [2nd] [◄] **1**	
[2nd] [INS] **00**	100→NTH:1/2→A:seq(A^N ,N,1,NTH,1)→LIST:sum LIST
[ENTER]	1

The TI-85 Menus

To leave the keyboard uncluttered, the TI-85 uses display menus to access many additional operations. The five keys immediately below the display are used to select items from menus. Specific menus are described in the appropriate chapters.

The Menu Keys

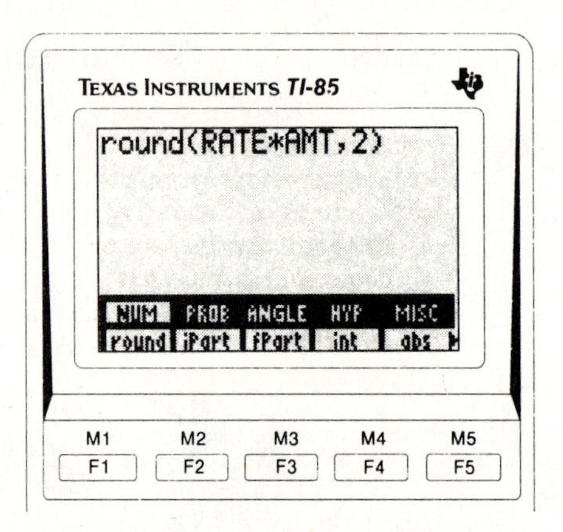

On the TI–85 keyboard, the menu keys are [F1], [F2], [F3], [F4], and [F5]. The 2nd operations of the menu keys are [M1], [M2], [M3], [M4], and [M5]. Menu items are shown above the five menu keys.

The Menu Items

Menu items can display on the bottom two lines (seventh and eighth lines) of the display. If any text is displayed on a line where a menu is to be displayed, the text in the display scrolls up a line.

The appearance of a menu item generally helps to identify what the menu item is.

- The names of functions, which return a value and are valid within an expression, generally begin with a lowercase letter; for example, **fPart** or **imag**.
- The names of instructions, which initiate an action from a command line, generally begin with a capital letter; for example, **Shade** or **ClDrw**.
- Menu items that access a lower-level menu or that perform immediate actions, generally are in all uppercase letters; for example, NUM or ZOUT.

Displaying Menus

If you select a menu item that displays another menu, the first menu may move to the seventh line; the new menu displays on the eighth line.

Displaying a Menu

Many of the 2nd operations, such as MATRX, VECTR, CPLX, MATH, and LIST, access menus of characters or names of variables, functions, and instructions to copy to the cursor location. When you press one of these keys, the eighth line of the display shows the menu items. For example, [2nd] [CPLX] labels the menu keys with complex number functions:

conj real imag abs angle

The menu items may access lower-level menus. For example, if you press [2nd] [MATH], the menu keys are labeled with the names of menus, each of which accesses a menu of math functions:

NUM PROB ANGLE HYP MISC

Displaying Additional Items in a Menu

A menu may have up to fifteen menu items, but only five are displayed at one time. ► at the right of the menu items indicates that there are more items in the menu. Press [MORE] to label the menu keys with the next group of menu items. If you are on the final group, [MORE] displays the first group. For example, on the MATH NUM menu:

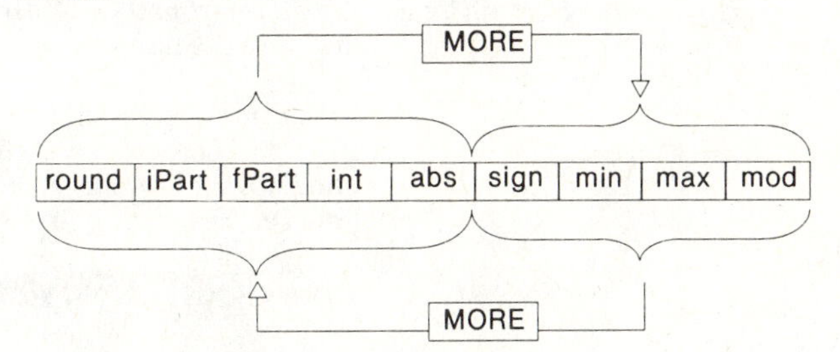

In this guidebook, all items in a menu usually are shown at once, stacked vertically; for example:

round iPart fPart int abs
sign min max mod

Selecting from Menus

You can select an item from the menu on the eighth line or from the menu on the seventh line. In this guidebook, menu items shown surrounded by brackets (for example, ⟨HYP⟩) indicate that you are to select that menu item.

Selecting an Item from the Menu on the Eighth Line

To select a menu item from the eighth line, press the corresponding menu key, [F1], . . . , [F5].

- If the item is a character or a name, it is copied to the cursor location, typing over existing characters (except in insert mode). If not all characters in a name can display, the name is truncated in the menu item, but the full name is copied to the cursor location. The menus do not change.
- If the item is an editing operation, such as INSr (insert row), the display changes as soon as you select the operation. The menus do not change.
- If the item is an action, such as SOLVE, the action occurs immediately. The menus change if appropriate.
- If the item accesses another menu, the menu keys are labeled immediately with the new menu.

The Menu on the Seventh Line

If you select a menu item that accesses another menu, the menu from the eighth-line may move to the seventh line, and the name of the selected menu is highlighted.

For example, selecting ⟨NUM⟩ from the MATH menu on the Home screen moves the MATH menu to the seventh line and displays the MATH NUM menu items in the eighth line. On the seventh line, NUM is highlighted.

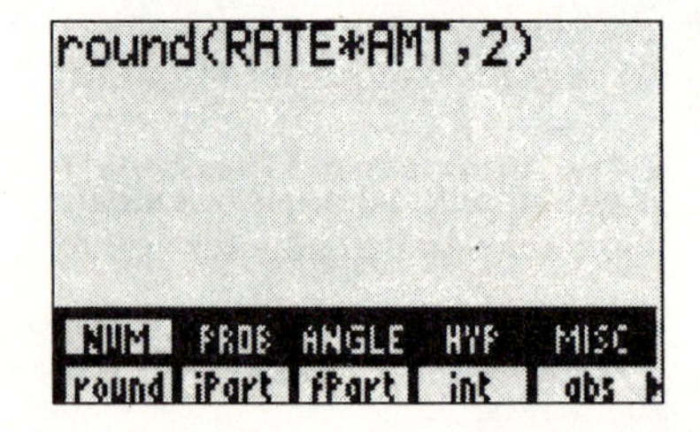

Accessing Menus from an Editor

An exception occurs if you are in a full-screen editor, such as the program or matrix editor. In this case, the editor menu remains on the seventh line for convenience.

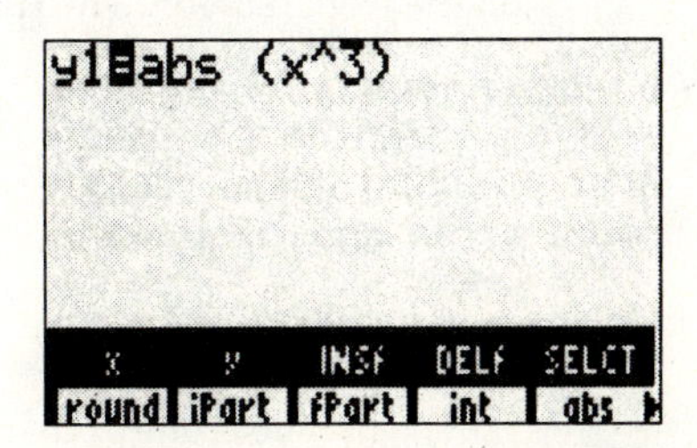

Selecting an Item from the Menu on the Seventh Line

If a menu is displayed on the seventh line, you can select an item from it in one of the following ways:

- Press [2nd] and then press the menu key, [M1], . . . , [M5], that corresponds to the item that you want. For example, [2nd] [M2] on the screen above would copy **y** to the cursor location.

- Press [EXIT], which causes the menu on the seventh line to "move down" to the eighth line. Then press the menu key ([F1], . . . , [F5]) that corresponds to the item that you want. For example, [EXIT] [F4] on the screen above would delete function **y1**.

"Exiting" a Menu

When you press [EXIT]:

- If a menu is displayed on the seventh line, that menu "moves down" to the eighth line. The display does not change.

- If a menu is displayed only on the eighth line, you are returned to the Home screen.

Moving around the TI-85

In addition to changes in the menu lines, the display may change when you press a key or select from a menu.

Moving to a Full-Screen Editor

Many of the keys on the TI–85 access applications with full-screen editors where you enter expressions as you do on the Home screen. The full-screen editors are:

CONS EDIT	POLY	GRAPH y(x)=
LIST EDIT	SOLVER	GRAPH r(θ)=
MATRX EDIT	SIMULT	GRAPH E(t)=
VECTR EDIT	MATH INTER	GRAPH Q′(t)=
STAT EDIT	STAT FCST	GRAPH RANGE
PRGM EDIT		GRAPH ZOOM ZFACT

When you select one of these:

- You "leave" the Home screen or the application in which you are working, and the appropriate editor displays.
- Any existing menu lines are cleared. The editor menu, if any, displays on the eighth line.

Working on a Full-Screen Editor

When you are working on a full-screen editor and press a key that displays a menu:

- The editor remains unchanged.
- The editor menu moves to the seventh line (if it is not already there), and the selected menu displays on the eighth line. You still can access editing operations (such as INSr) or instructions (such as SOLVE) with the [2nd] key.

Leaving an Editor

To leave an editor:

- Press [2nd] [QUIT] to return to the Home screen.
- Press [EXIT] one or more times to return to the previous menu, display, or the Home screen.
- Press the appropriate keys to move to another application, such as [2nd] [SOLVER].

Pull-Down Screens

The VARS and CATALOG selection screens temporarily replace the current display.

- The current display is replaced, but you have not "left" the application in which you are working.
- The VARS or CATALOG menu is displayed.

When you press [EXIT] or make a selection, the current display and menus are shown again.

The Prompt Line

Sometimes you will be prompted for a value or variable name on the prompt line, the line above the menu(s).

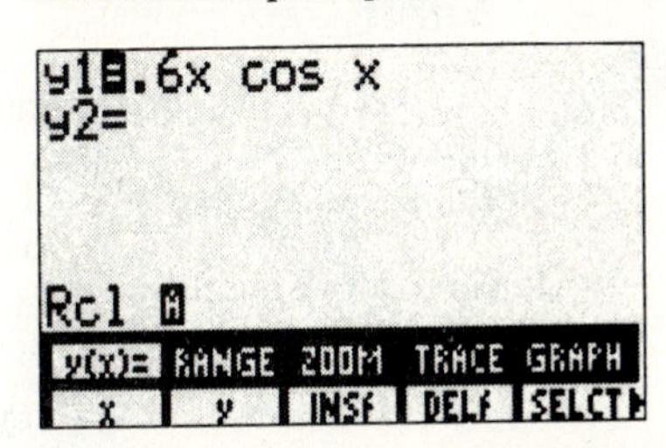

Clearing a Prompt

Press [CLEAR] to clear anything on the prompt line. Press [CLEAR] on a blank prompt line to clear the prompt and return the cursor to the editor or graph.

Correcting an Error on the Prompt Line

When an error occurs on the prompt line, ERR *nn* is displayed at the right of the line. It is not necessary to clear the error message to edit the entry. To clear the error and the entry, press [CLEAR].

Returning to the Home Screen

To return to the Home screen from any other screen, press [2nd] [QUIT].

You also can press [EXIT] one or more times until the Home screen is displayed.

The CATALOG

You can use the CATALOG to copy the name of an instruction or a function to the cursor location in an expression that you are editing. These include the functions and instructions from the keyboard and from menus.

The CATALOG Selection Screen

When you press [2nd] [CATALOG], the CATALOG screen temporarily replaces the screen on which you are working.

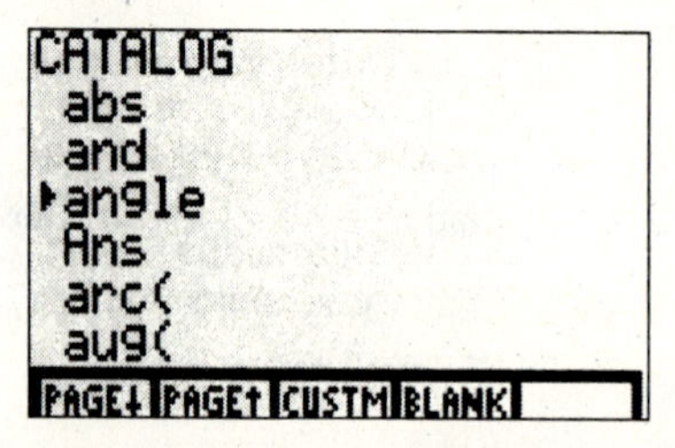

The names of functions and instructions are displayed in alphabetical order. Names that do not begin with an alphabetic character (such as **+** or **►Bin**) follow **Z**. An arrow at the left of the name indicates the selection cursor. To move around the list:

- Press a letter to move quickly to names beginning with that letter. (The keyboard is set in ALPHA-lock.) Uppercase and lowercase names are intermixed.
- Press [▲] on the first item in the CATALOG to move quickly to names beginning with special characters at the end of the list.
- Use ‹PAGE↓› and ‹PAGE↑› to move to the next page of names.
- Use [▼] and [▲] to move down and up the list.

Copying a Name to an Expression

Press [ENTER] to select the name to copy. The CATALOG selection screen disappears and the name is copied to the cursor location.

Leaving the CATALOG

To leave the CATALOG without making a selection:

- Press [EXIT] or [CLEAR] to return to the application in which you are working.
- Press [2nd] [QUIT] to return to the Home Screen.

Defining and Editing Matrices with the Editor

In addition to entering matrices directly in an expression, you can use the matrix editor to define a new matrix or to edit an existing matrix. To define a new matrix or edit an existing one, you must first select the matrix name.

Selecting a Matrix

1. Select ⟨EDIT⟩ from the MATRX menu to display the matrix selection screen. The menu keys are labeled with the names of existing matrices in alphabetical order.

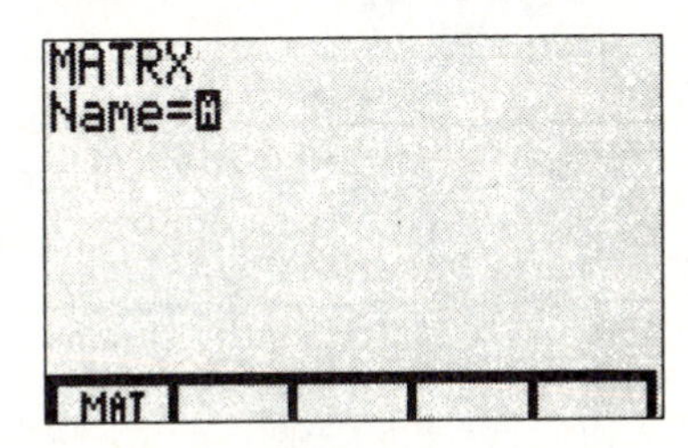

2. Enter the name of the matrix.

   - Select an existing name from the menu.

   - Type the name of a new or existing matrix of up to eight characters (case-sensitive). The keyboard is set in ALPHA-lock.

3. Press [ENTER]. If you selected an existing matrix, its dimensions and elements are displayed.

Accepting or Changing Matrix Dimensions

The dimensions of the matrix (rows × columns) are displayed on the top line. The default dimension for a new matrix is 1 × 1. The cursor is on the row dimension. You must accept or change the row dimension value and the column dimension value each time you enter the matrix editor.

- To accept the value, press [ENTER].

- To change the value, enter a number (up to 255) and press [ENTER].

Note: You can use [▲] and [▼] to move onto and edit the matrix dimensions at any time in the editor.

Displaying Matrix Contents in the Editor

The matrix is displayed in the matrix editor one column at a time. For example, let **SAMPLE** be the 8 x 4 matrix:

$$\begin{bmatrix} 1 & 2 & 1 & 4 \\ 2 & 2 & 2 & 2 \\ 1 & 3 & 3 & 4 \\ 0 & 0 & 5 & 3 \\ 2 & 0 & 9 & 4 \\ 5 & 8 & 0 & 0 \\ 5 & 0 & 2 & -4 \\ 5 & 6 & 3 & 1.1 \end{bmatrix}$$

The six elements indicated in column 3 of **SAMPLE** would be displayed in the matrix editor as:

```
MATRX:SAMPLE    8x4
↑2,3=2
 3,3=3
 4,3=5
 5,3=9
 6,3=0
↓7,3=2
◂COL  COL▸  INSr  DELr  INSc ▸
```

Editing a Matrix with the Matrix Editor

In a new matrix, all values are zero. ↓ is displayed at the left of the line above the menu(s) if there are more rows in the matrix than can be displayed at one time.

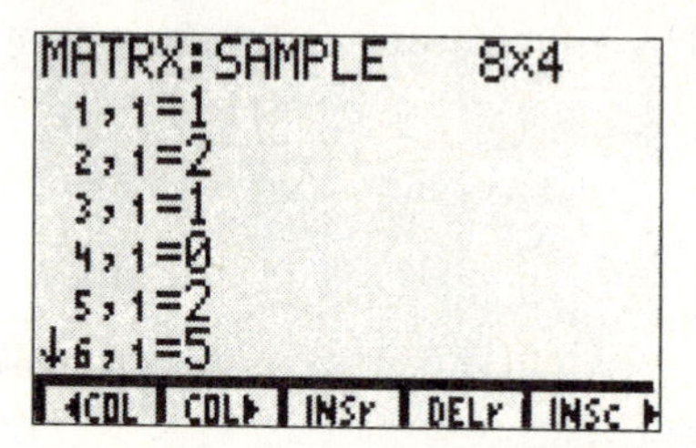

name & row × col
row, col = value

Enter new real or complex values (which can be expressions) for the matrix elements, as appropriate. Expressions are evaluated when you move off the element or leave the editor.

- Press [ENTER] after each value to enter the matrix row by row.
- Press [▼] after each value to enter the matrix column by column.

Note: If you press a key that accesses a menu, the matrix editor menu moves to the seventh line (if it is not already there), and the selected menu is displayed on the eighth line.

Entering SIMULT (Simultaneous) Equations

[2nd] [SIMULT] accesses the Simult (simultaneous) Equations solving capabilities of the calculator. You can solve systems of up to 30 linear equations with 30 unknowns.

Entering the Equations

1. Press [2nd] [SIMULT]. The SIMULT screen appears.

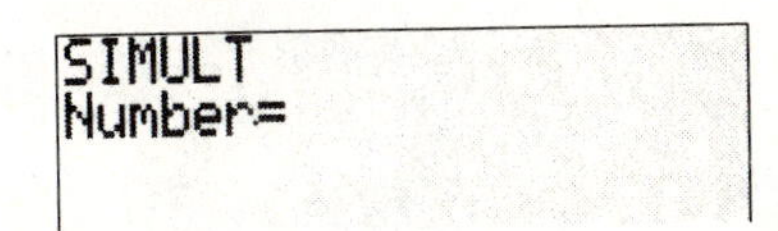

2. Enter an integer between 2 and 30 (which can be an expression) for the number of simultaneous equations. Press [ENTER]. The coefficient entry screen for the first equation appears. An example for a system of four equations and four unknowns is shown. The equation is displayed on the top line for reference; you cannot edit it.

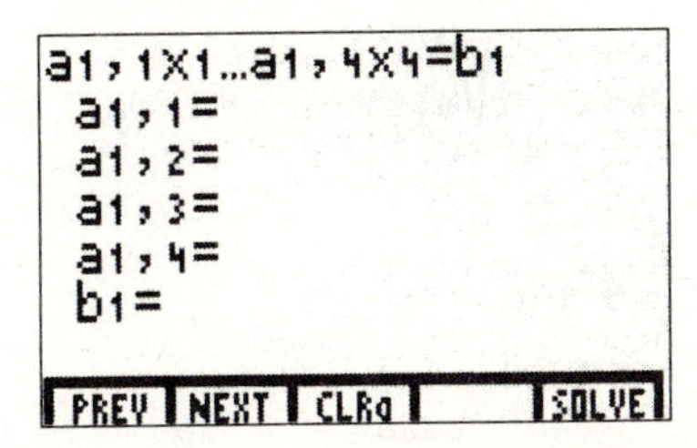

3. Enter a real or complex value (which can be an expression) for the first coefficient, $a_{1,1}$. Press [ENTER].

4. Enter all coefficients for the first equation. If you press [ENTER] after entering the last coefficient or select ⟨NEXT⟩, the second equation is displayed. Enter the remaining coefficients.

⟨PREV⟩ and ⟨NEXT⟩ move between equations. [▲], [▼], and [ENTER] move between coefficients and equations. [CLEAR] clears only the line on which the cursor is located. ⟨CLRa⟩ clears the coefficients for the current equation.

Note: If you press a key that accesses a menu, the SIMULT editor menu moves to the seventh line (if it is not already there), and the selected menu is displayed on the eighth line.

Solving Simultaneous Equations

After you find the solutions to the simultaneous equations, you can store the results.

Solving the Equations

After entering the coefficients, select 〈SOLVE〉.

Storing the Coefficients or Results

The results are displayed only; they cannot be edited and they are not stored in memory. The coefficients are used for SIMULT entry only; they do not update variables a11, b1, x1, etc.

- To store coefficients $a_{1,1}$, $a_{1,2}$, ..., $a_{n,n}$ into an $n \times n$ matrix, select 〈STOa〉.
- To store coefficients b_1, b_2, ..., b_n into a vector of dimension n, select 〈STOb〉.
- To store the results x_1, x_2, ..., x_3 into a vector of dimension n, select 〈STOx〉.

Storing a Single Value

You can store any value on the coefficients entry or results screen to a variable. Press [STO►] and enter the variable name after the Name= prompt.

Editing the Equation

You can edit the coefficients and calculate new solutions. Select 〈COEFS〉 to return to the first coefficient entry screen.

The simult Function in an Expression

The **simult** function on the Home screen or in a program, which can be copied from the CATALOG, accesses the SIMULT equation-solver feature.

simult(*a_matrix*,*b_vector***)**

a_matrix is an n×n real or complex matrix containing the *a* coefficients. *b_vector* is an n-dimension real or complex vector containing the *b* coefficients. When the expression is evaluated, the result is an n-dimension vector containing the values of x.

Example: Simultaneous Equations

The SIMULT feature of the TI-85 can solve large systems of linear equations. Solve the 10 by 10 system below.

Problem

$$
\begin{aligned}
4x_1+9x_2+7x_3+8x_4+3x_5+5x_6+3x_7+5x_8+8x_9+6x_{10} &= 3\\
8x_1+3x_2+8x_3+9x_4+9x_5+5x_6+4x_7+7x_8+0x_9+0x_{10} &= 7\\
1x_1+2x_2+6x_3+7x_4+7x_5+0x_6+3x_7+4x_8+1x_9+5x_{10} &= 9\\
4x_1+4x_2+0x_3+3x_4+0x_5+5x_6+7x_7+7x_8+2x_9+4x_{10} &= 6\\
7x_1+5x_2+0x_3+7x_4+0x_5+9x_6+3x_7+6x_8+1x_9+0x_{10} &= 5\\
2x_1+7x_2+0x_3+3x_4+4x_5+7x_6+8x_7+8x_8+3x_9+9x_{10} &= 1\\
2x_1+6x_2+1x_3+5x_4+2x_5+4x_6+7x_7+8x_8+4x_9+7x_{10} &= 5\\
4x_1+3x_2+6x_3+7x_4+0x_5+7x_6+9x_7+1x_8+6x_9+4x_{10} &= 0\\
2x_1+1x_2+9x_3+3x_4+8x_5+6x_6+9x_7+5x_8+7x_9+5x_{10} &= 0\\
9x_1+4x_2+3x_3+0x_4+9x_5+3x_6+8x_7+0x_8+1x_9+1x_{10} &= 0
\end{aligned}
$$

Procedure

1. Press [2nd] [SIMULT]. Enter **10** for the number of equations.

2. Enter the coefficients for each of the equations in the coefficient editor.

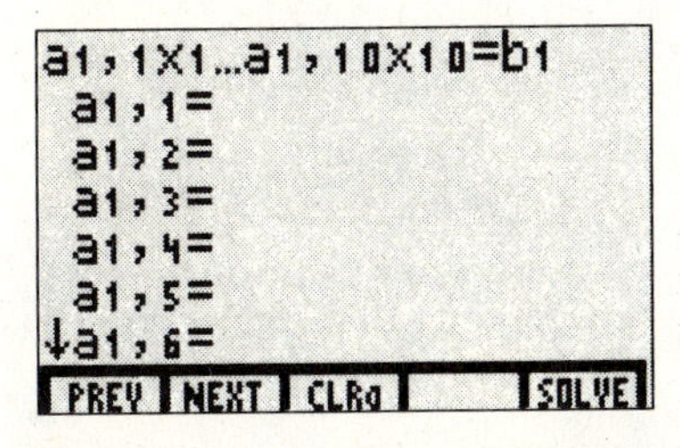

3. Select ⟨SOLVE⟩. The results are displayed.

4. Select ⟨STOa⟩, ⟨STOb⟩, and ⟨STOx⟩ to store the coefficients and results to **SA**, **SB**, and **SX**.

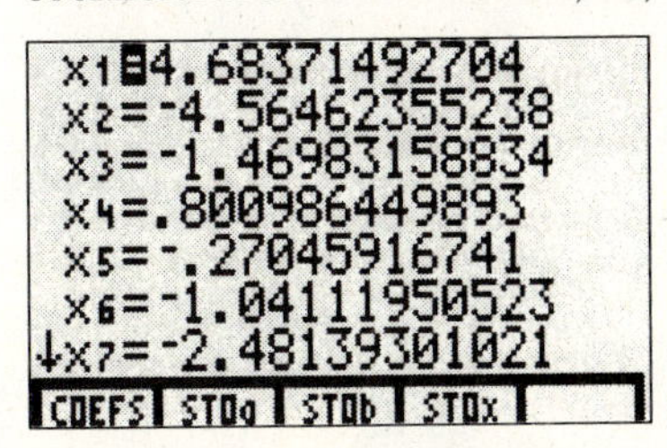

APPENDICES

APPENDIX A

A Guide to the Variables of Earth Algebra

For quick reference, each of the variables used to define a function in *Earth Algebra* is listed here with its definition. These include only the variables specifically used in the studies of environmental topics, not those which appear in the "prerequisite" chapters. The variables are listed in the order in which they appear in the text, and the page number refers to the page on which the variable is first introduced.

CO_2C = atmospheric CO_2 concentration (in ppm) by year, page 34 (linear model).

GT = average global temperature increase (in degrees Fahrenheit), page 47.

OL = average ocean level increase (in feet), page 49.

A = number ($\times 10^6$) of automobiles in the United States by year, page 89.

MPG = average fuel efficiency (miles per gallon) per automobile in the United States by year, page 91.

GPM = gallons of gasoline burned per mile by year, page 98.

M = average number of miles ($\times 10^3$) each automobile in the United States drives per year, page 100.

CC = coal consumption (in quads) in the United States by year, page 106.

PC = petroleum consumption (in quads) in the United States by year, page 107.

NGC = natural gas consumption (in quads) in the United States by year, page 109.

TEC = total energy consumption (in quads) in the United States by year, page 110.

L = number of hectares ($\times 10^6$) of rain forest cut for logging by year, page 126.

CG = number of hectares ($\times 10^6$) of rain forest cut for cattle grazing by year, page 127.

AD = number of hectares ($\times 10^6$) of rain forest cut for agriculture and development by year, page 132.

CO_2A = total CO_2 emission (in lb. $\times 10^9$) due to automobiles in the United States by year, page 134.

TCE = total carbon emission (in gigatons) due to energy (from coal, natural gas, and petroleum) consumption in the United States by year, page 136.

CDF = total carbon emission (in metric tons) due to deforestation by year, page 137.

CO_2 = atmospheric CO_2 concentration (in ppm) by year, page 148 (exponential model).

C = worldwide carbon emission (in gigatons) by year, page 153.

AC = total atmospheric carbon accumulation (in gigatons) by year, page 153.

ppm = parts per million of atmospheric carbon dioxide, used to convert gigatons of atmospheric carbon to ppm, page 154.

P = United States population ($\times 10^6$) by year, page 175.

GNP = United States gross national product (in dollars $\times 10^{12}$) by year, page 183.

EC = total carbon emission (in gigatons) due to energy consumption in the United States by year, page 212.

OC = total carbon emission (in gigatons) due to all sources in the United States other than energy consumption, page 212.

TE = target carbon emission (in gigatons) from energy consumption in the United States in future years, page 213.

APPENDIX B

A Short Table of Conversions

1 hectare = 2.47 acres

1 ton = 2000 lbs.

1 gigaton = 10^9 metric tons

1 metric ton = 1000 kilograms

1 lb. = .454 kilograms

1 quad = 10^{15} BTU

weight CO_2 = weight carbon × 3.667

1 gallon of gasoline burned emits 20 lbs. CO_2

1 quad of coal burned emits .02500 gigatons of carbon

1 quad of natural gas emits .01454 gigatons of carbon

1 quad of petroleum emits .02045 gigatons of carbon

1 hectare of destroyed rain forest emits approximately $\frac{1}{2}$ ton of carbon

APPENDIX C

The Quadratic Formula and Complex Numbers

The quadratic formula provides all solutions to quadratic equations. Its derivation is relatively simple using the method of completing the square. Consider the standard form for a quadratic equation,

$$ax^2 + bx + c = 0.$$

First, divide by the coefficient of x^2 to get

$$x^2 + \frac{b}{a}x + \frac{c}{a} = 0;$$

transfer the constant $\frac{c}{a}$ to the right to get

$$x^2 + \frac{b}{a}x = -\frac{c}{a}.$$

Now, completing the square on the left side of this equation means adding an appropriate term so that the result factors as a perfect square, i.e, has the form $(x - h)^2$. This appropriate term is obtained by squaring half of $\frac{b}{a}$; i.e., add $\left(\frac{b}{2a}\right)^2$ to the left, and of course, to the right side too. This gives

$$x^2 + \frac{b}{a} + \frac{b^2}{4a^2} = \frac{b^2}{4a^2} - \frac{c}{a}.$$

Simplify the right and factor the left to get

$$\left(x+\frac{b}{2a}\right)^2=\frac{b^2-4ac}{4a^2}.$$

Remember, we are solving the equation for x, so now take the square root of both sides:

$$x+\frac{b}{2a}=\pm\frac{\sqrt{b^2-4ac}}{2a};$$

finally solve for x and simplify to get the celebrated *quadratic formula*,

$$x=\frac{-b\pm\sqrt{b^2-4ac}}{2a}.$$

The quantity b^2-4ac, which is under the radical, is significant. If $b^2-4ac>0$, there are two real solutions; if $b^2-4ac=0$, there is only one real solution; if $b^2-4ac<0$, there is no real solution.

The method of completing the square can be used to solve any quadratic equation should you forget the quadratic formula, or if it seems easier.

We discuss the special case where $b^2-4ac<0$ in more detail. As pointed out earlier, there will be no real solution to the quadratic equation. However, there is a type of number, called a complex number (sometimes known as imaginary), which will be a solution. Define

$$\sqrt{-1}=i,$$

and a *complex number* to be one which can be written in the form $\alpha+\beta i$, where α and β are real numbers.

The next two examples illustrate how complex numbers are solutions to quadratic equations.

EXAMPLE 1

Solve $x^2+1=0$.

This is easier without using the quadratic formula. Simply transfer the 1 to the right to get

$$x^2=-1$$

$$x=\pm\sqrt{-1},$$

so

$$x = \pm i.$$

Thus the equation $x^2 + 1 = 0$ has no real solution, but has two complex solutions.

EXAMPLE 2

Solve $x^2 + 2x + 2 = 0$.

The quadratic formula gives

$$x = \frac{-2 \pm \sqrt{-4}}{2}.$$

To simplify, write $\sqrt{-4} = \sqrt{4}\sqrt{-1} = 2i$, so

$$x = \frac{-2 \pm 2i}{2} = 1 \pm i.$$

We complete this appendix with a brief discussion of operations on complex numbers. They can be added, subtracted, multiplied and divided according to the following rules.

1. Addition: $(\alpha + \beta i) + (\gamma + \delta i) = (\alpha + \gamma) + (\beta + \delta)i$.
2. Subtraction: $(\alpha + \beta i) - (\gamma + \delta i) = (\alpha - \gamma) + (\beta - \delta)i$.
3. Multiplication: $(\alpha + \beta i)(\gamma + \delta i) = \alpha\gamma + \beta\gamma i + \alpha\delta i + \beta\delta i^2$

   $$= (\alpha\gamma - \beta\delta) + (\beta\gamma + \alpha\delta)i,$$

   or use FOIL and simplify.
4. Division. This requires a little work. The *conjugate* of the complex number $\alpha + \beta i$ is $\alpha - \beta i$. One significance of the conjugate is that the product of $\alpha + \beta i$ and its conjugate is a non-zero real number (provided $\alpha + \beta i \neq 0$):

   $$(\alpha + \beta i)(\alpha - \beta i) = \alpha^2 + \beta^2.$$

Back to division of complex numbers:

$$\frac{\gamma + \delta i}{\alpha + \beta i} = \frac{(\gamma + \delta i)(\alpha - \beta i)}{(\alpha + \beta i)(\alpha - \beta i)} = \frac{(\alpha\gamma + \beta\delta) + (\alpha\delta - \beta\gamma)i}{\alpha^2 + \beta^2} = \frac{\alpha\gamma + \beta\delta}{\alpha^2 + \beta^2} + \frac{\alpha\delta - \beta\gamma}{\alpha^2 + \beta^2}i.$$

Thus division is accomplished by multiplying numerator and denominator by the conjugate of the denominator, thus converting the denominator to a real number.

Here is an example of each of these operations:

1. Addition: $(3 - 2i) + (-1 + 5i) = 2 + 3i$;

2. Subtraction: $(7 - 6i) - (3 - i) = 4 - 5i$;

3. Multiplication:
$$\begin{aligned}(2 + 3i)(-1 - 4i) &= -2 - 3i - 8i - 12i^2 \\ &= -2 - 11i + 12 \\ &= 10 - 11i;\end{aligned}$$

4. Division: $\dfrac{1-2i}{3+4i} = \dfrac{(1-2i)(3-4i)}{(3+4i)(3-4i)} = \dfrac{-5-10i}{9+16} = \dfrac{-5}{25} - \dfrac{10}{25}i = -\dfrac{1}{5} - \dfrac{2}{5}i.$

This completes the real numbers.

APPENDIX D

Answers to Odd-Numbered Exercises in Things to Do Sections

TO THE STUDENT

If you need further help, you may want to obtain a copy of the *Student's Solution Manual* that goes with this book. It contains solutions to all the odd-numbered *Things to Do* exercises. Your college bookstore either has this book or can order it for you. To order, use ISBN 0-06-501555-X.

The answers provided in the text, as well as supplemental material, are the ones we think students will obtain when they work the exercises using the methods explained in the text. However, in many cases there are equivalent forms of the answer that are correct. For example, if the answer sections shows $\frac{1}{4}$ and you obtain .25, then you have obtained the right answer but written in a different (but equivalent) form. Sometimes when and where you round off numbers can change an answer somewhat.

In general, if your answer does not agree with the one given in the text, see whether it can be transformed into another form. If you still have doubts, talk to your instructor.

CHAPTER 1

SECTION 1.4

1. $f(2) = 1; f(-3) = 11; f(1.01) = 2.98;$

$f\left(-\frac{1}{2}\right) = 6;$

Domain = all real numbers.

3. $h(2) = -\frac{7}{9}; h(-3) = -\frac{22}{29};$

$h(1.01) = .1497;$

$h\left(-\frac{1}{2}\right) = \frac{17}{76};$

Domain = all real numbers $x \neq \frac{7}{4}$.

5. $T(2.4) = -119.988; T(24) = -119.88;$

$T(240) = -118.8; T(2400) = -108;$

$T(24000) = 0;$

Domain = all real numbers;

7. $H(2) = -1; H(-3.3) = -1.57;$

$H(11.001) = 32.55; H(5.5) =$ undefined;

Domain = all real numbers $x \neq \frac{11}{2}$.

9. $R(0) = \sqrt{63}; R(-2) = 8.198;$

$R(30) = 0; R(-1.52) = 8.136;$

$R(50) = \sqrt{-42}$, which is undefined;

Domain = all real numbers $s \leq 30$.

CHAPTER 2

SECTION 2.6

1. $y = -1.4x$; slope $= -1.4$, y-intercept $= (0, 0)$.

3. $y = \left(\frac{2}{3}\right)x - \frac{29}{3}$; slope $= \frac{2}{3}$,

y-intercept $= \left(0, \frac{-29}{3}\right)$.

5. $y = x + \frac{8}{3}$; slope $= 1$,

y-intercept $= \left(0, \frac{8}{3}\right)$.

7.

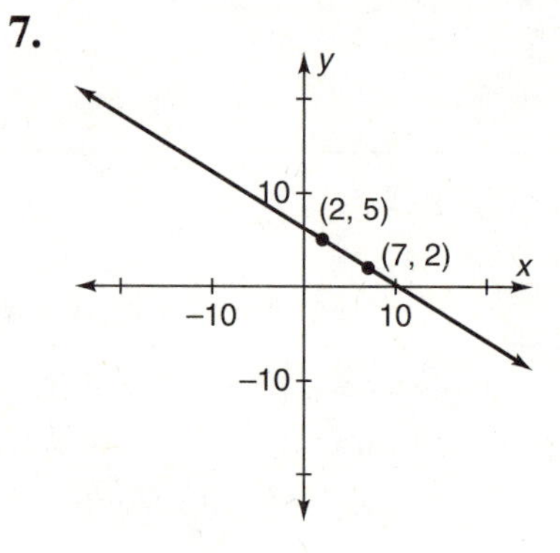

9.

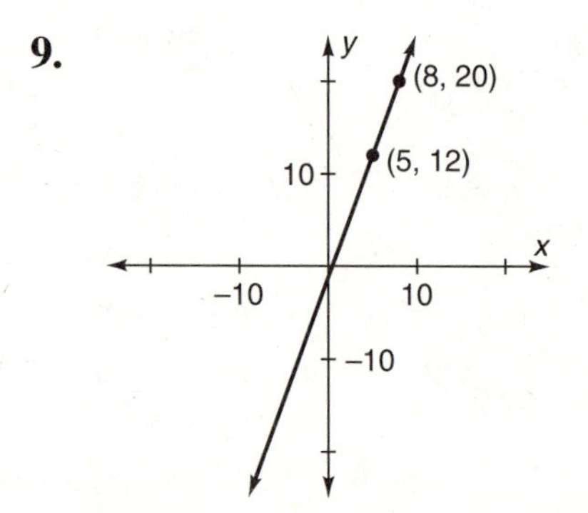

11. $m = 2.4$

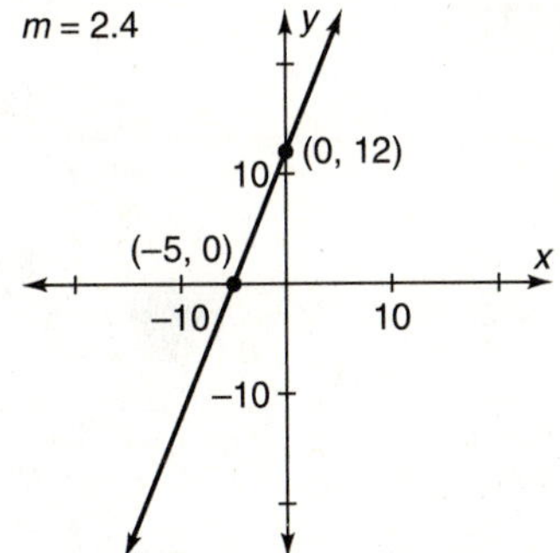

13. $m = .0003$

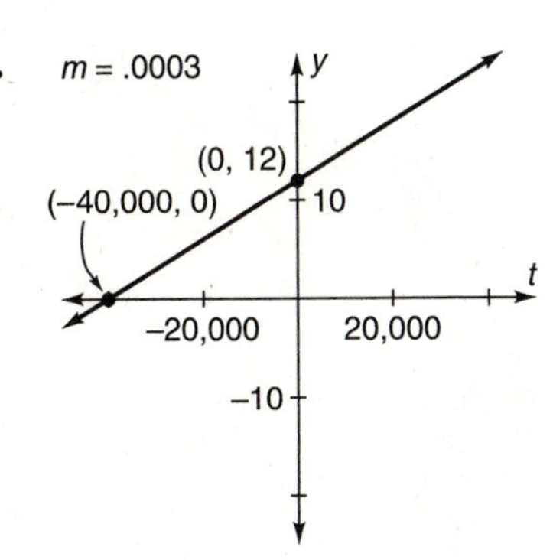

15. $m = 5.8$

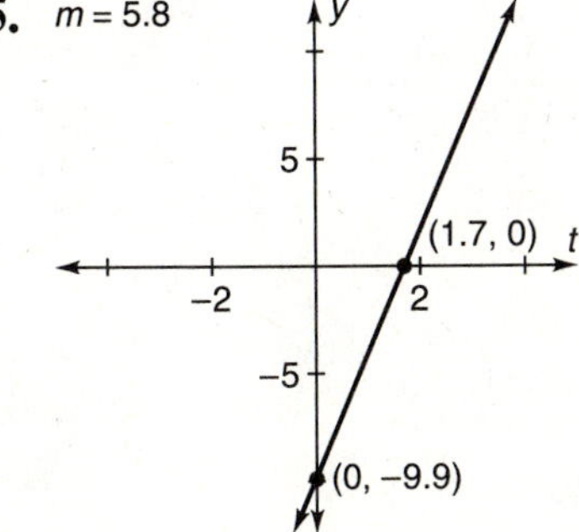

17. $y = .92x + 33.1$; slope $= .92$,

y-intercept $= (0, 33.1)$,

x-intercept $= (-35.98, 0)$.

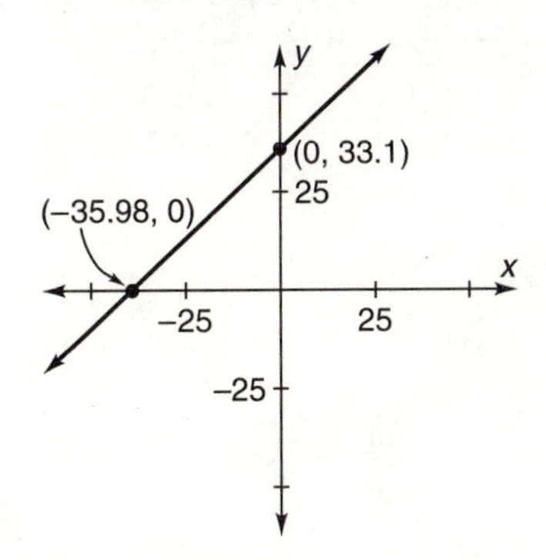

19. $y = -.006x + 2.8$; slope $= -.006$,

y-intercept $= (0, 2.8)$,

x-intercept $= (466.67, 0)$.

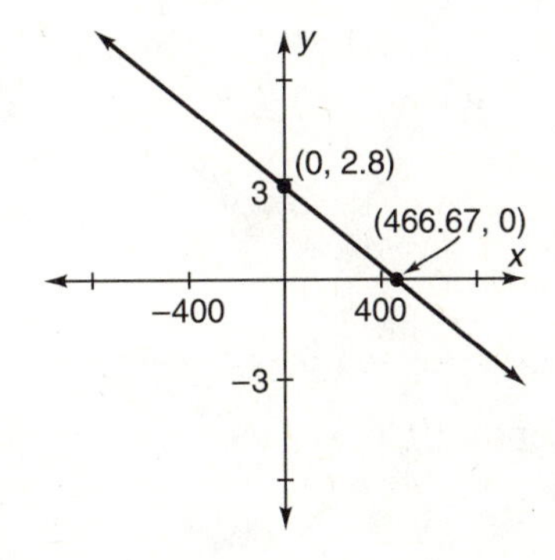

21. $y = -.02x + 23.8$; slope $= -.02$,

y-intercept $= (0, 23.8)$,

x-intercept $= (1190, 0)$.

(See page 232 for the graph to Exercise 21.)

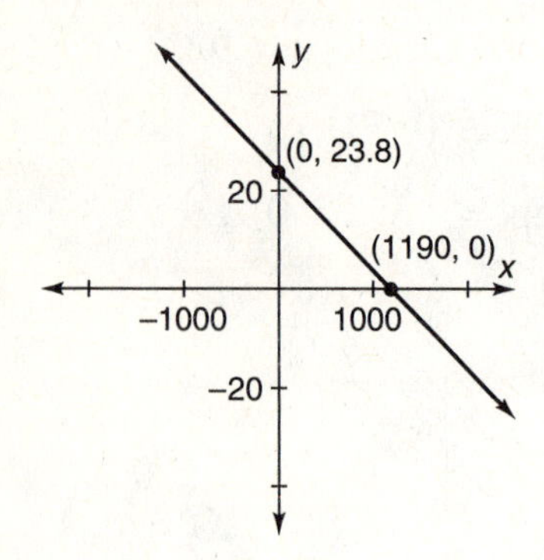

23. $y = \left(-\frac{5}{2}\right)x$; slope $= -\frac{5}{2}$,

y-intercept = (0, 0).

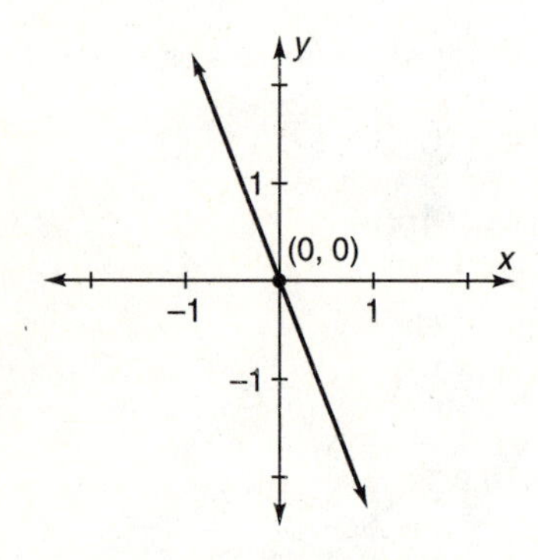

25. $y = -1304.35x + 6.57$; slope $= -1304.35$,

y-intercept = (0, 6.57),

x-intercept = (.005, 0).

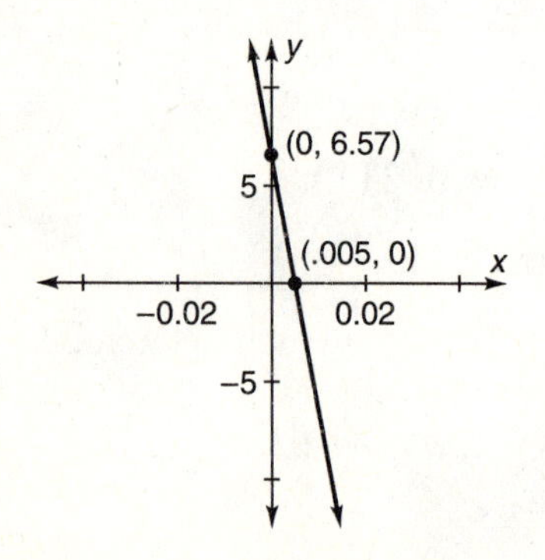

CHAPTER 4

SECTION 4.2

1. $f(g(x)) = 1 - 3x$; $g(f(x)) = 7 - 3x$.

3. $r(s(x)) = \dfrac{1}{5x+2}$; $s(r(x)) = \dfrac{5+2x}{x}$.

5. $E(H(x)) = .13578x^3 - .00023$;

$H(E(x)) = .13578x^3 - .00023$;

$E(H(0)) = -.00023$;

$E(H(1)) = .13555$;

$H(E(1)) = .13555$.

7. A) $P(F(w)) = 4 + .2w$;

B) $4.00;

C) 40 gallons;

D) 100 gallons.

CHAPTER 6

SECTION 6.7

1. Range = all $y \geq -4$.

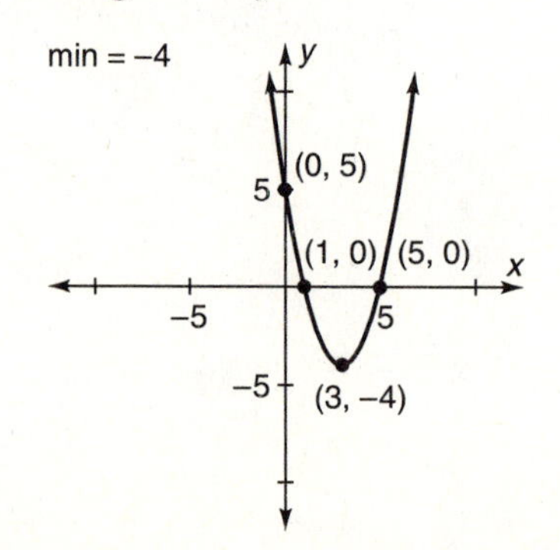

3. Range = all $y \geq -49,743.1$.

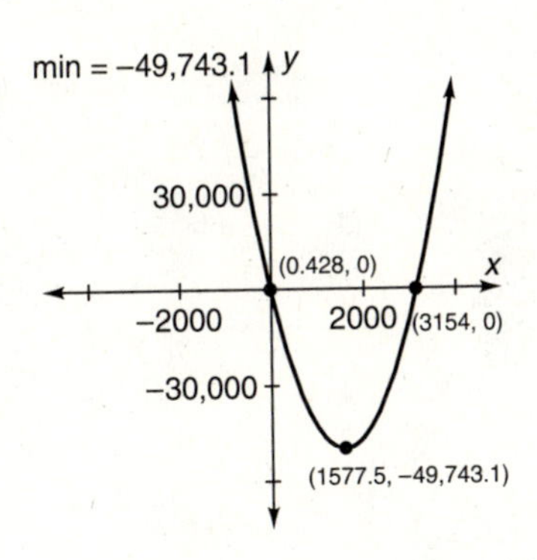

5. Range = all $y \leq 76.12$.

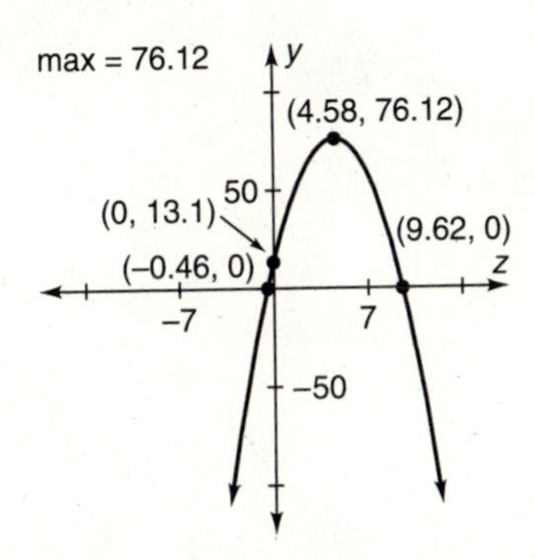

7. Range = all $y \leq 9.03$.

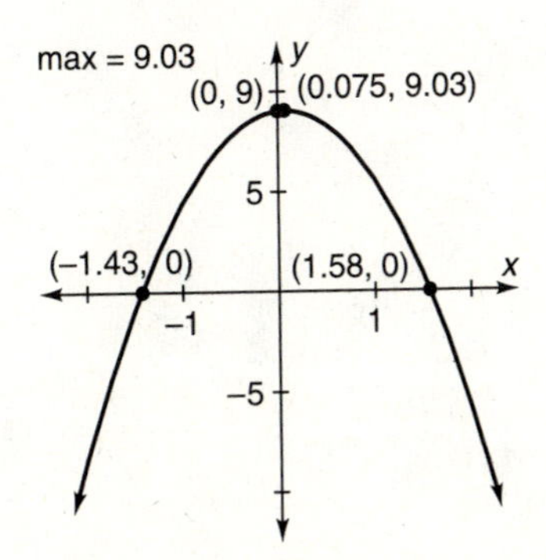

9. $x = 3.57$ or -4.77.

11. $x = 1.63$ or 7.53.

13. vertex = (105, 6.285).

15. vertex = (181.85, −77.81);

$K(b) = 45$ when $x = 383.993$ or -20.67;

$K(b) = 50$ when $x = 388.07$ or -24.73.

CHAPTER 7

SECTION 7.4

1. $x = 1, y = 0$.

3. $x = .2, y = -.8$.

5. $x = \frac{1}{2}, y = -1$.

7. $x = 2; y = 0; z = 1$.

9. No solution.

11. $\begin{bmatrix} 3 & 3 \\ 3 & 5 \\ 8 & 8 \end{bmatrix}$.

13. $\begin{bmatrix} -3 & 0 \\ 0 & 3 \end{bmatrix}$.

15. $\begin{bmatrix} 10 & 5 \\ 15 & 30 \end{bmatrix}$.

17. $\begin{bmatrix} 6 & -6 \\ 0 & -12 \end{bmatrix}$.

19. Not defined.

21. $\begin{bmatrix} 9 \\ 17 \end{bmatrix}$.

23. Not defined.

25. $\begin{bmatrix} 7 \\ 7 \end{bmatrix}$.

27. $\begin{bmatrix} 2 & 3 & 7 \\ 3 & 1 & 2 \\ 1 & 5 & 8 \end{bmatrix}$.

29. Not defined.

31. Yes.

33. $x = 1, y = 1$.

35. $x = .2, y = -.4$.

37. $x = 2.2, y = .7$.

39. $x = .5102, y = 4.3061$.

41. $x = 2.1, y = -2.3, z = -.8$.

43. $x = 4.28, y = -7.56, z = 3.2$.

45. $x = .0015, y = -0.85, z = 14.6$.

CHAPTER 10

SECTION 10.4

1. 1.816.

3. 7.389.

5. 141667020.5.

7.

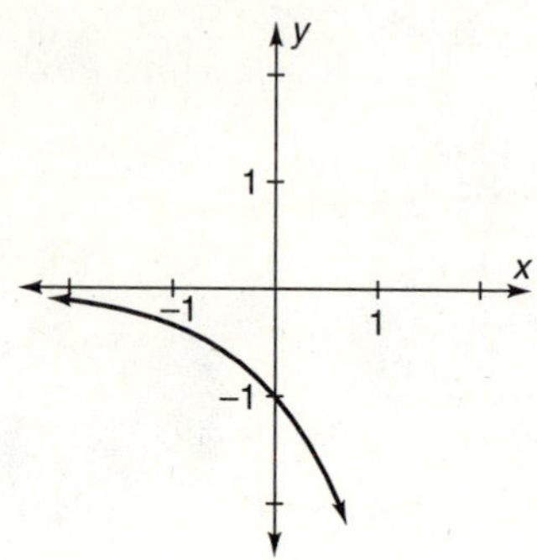

9.

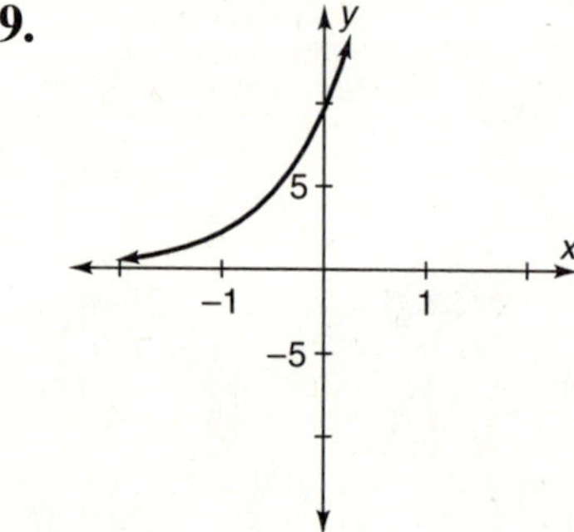

11. 4.

13. 4.

15. −1.6383.

17. −3.5066.

19. (316.23, 0).

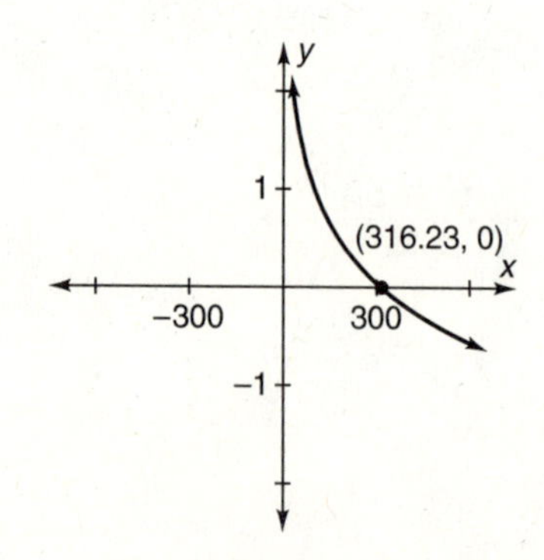

21. (1.3668, 0)

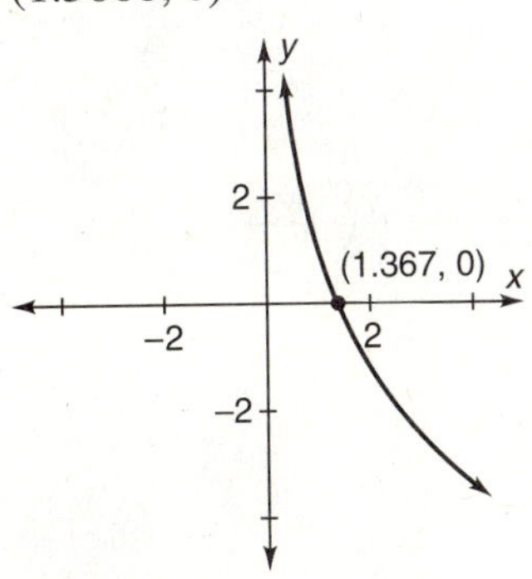

23. $8.2316 = t$

25. $.3377 = s$

27. $.5 = t$

CHAPTER 13

SECTION 13.2

1. $a = 1, r = 6, n = 7, S = 335,923.$

3. $a = 71.2,\ r = 4.5, n = 12,$
$S = 3{,}707{,}284{,}998$ (rounded).

5. $a = .5, r = .1, n = 793, S = .556.$

7. $4.1 + 4.1(9.6)^1 + 4.1(9.6)^2 + 4.1(9.6)^3 + 4.1(9.6)^4 + 4.1(9.6)^5 + 4.1(9.6)^6 + 4.1(9.6)^7;\ S = 34{,}391{,}828.29.$

CHAPTER 16

SECTION 16.5

1. $f^{-1}(y) = \frac{1}{2}y + \frac{1}{2}$

3. $g^{-1}(y) = 5y - 100$

5. $f^{-1}(y) = \frac{-y}{5.4}$

7. $h^{-1}(y) = \frac{\ln y - \ln 2.1}{\ln 7.01};$
$h(4.3) = 9095.085205;$
$h^{-1}(9095.085205) = 4.3.$

9. $g^{-1}(y) = \frac{\ln(-y) - \ln 3}{\ln 1.2};\ g(1) = -3.6;$
$g^{-1}(-3.6) = 1.$

11. $f^{-1}(y) = e^{\frac{y-2.7}{1.3}};\ f(2) = 3.601;$
$f^{-1}(3.601) = 2.$

13. $h^{-1}(y) = \frac{1}{3}e^{\frac{y+9}{1.5}};\ h(8) = -4.2329;$
$h^{-1}(-4.2329) = 8.$

15. $U^{-1}(y) = \frac{1}{2}e^{\frac{y}{.894}};\ U(-3) =$ no solution.

CHAPTER 19

SECTION 19.3

1.

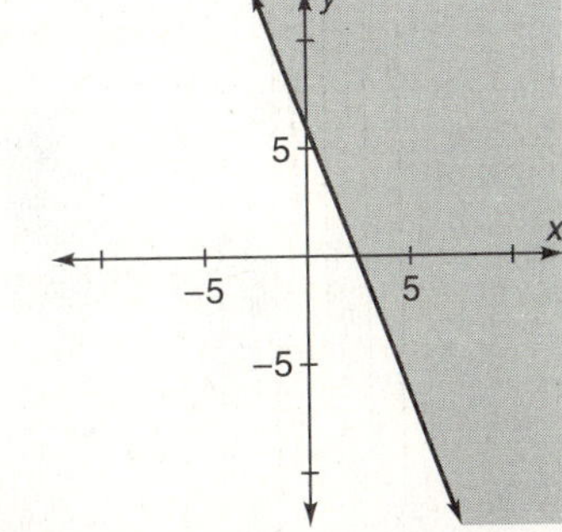

3.

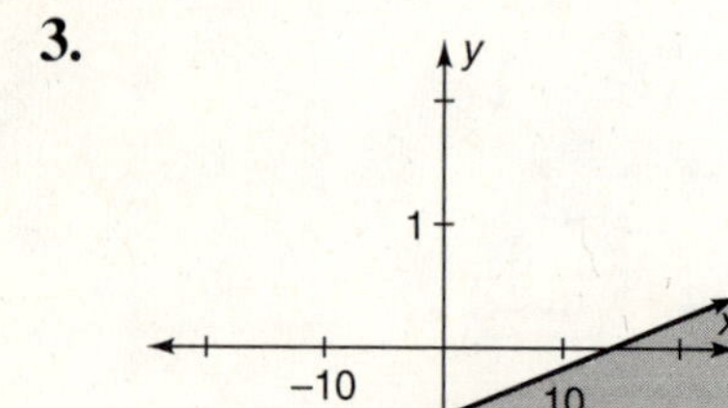
y
1
−10
10
x
−1

7.

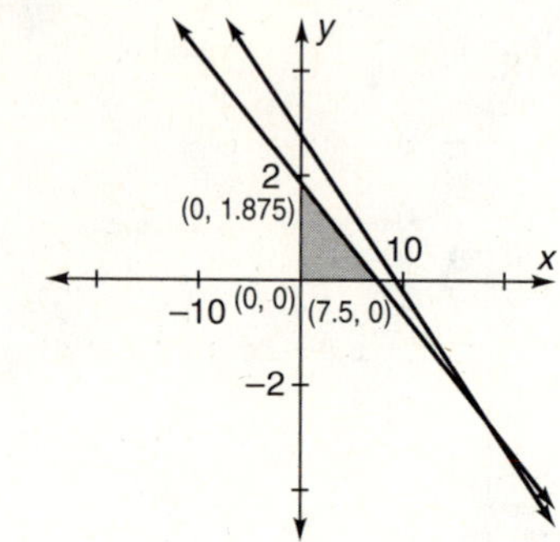
y
2
(0, 1.875)
10
x
−10
(0, 0)
(7.5, 0)
−2

5.

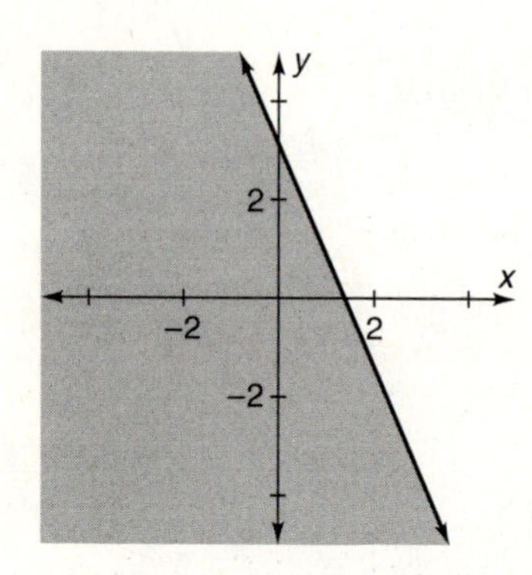
y
2
x
−2
2
−2

9.

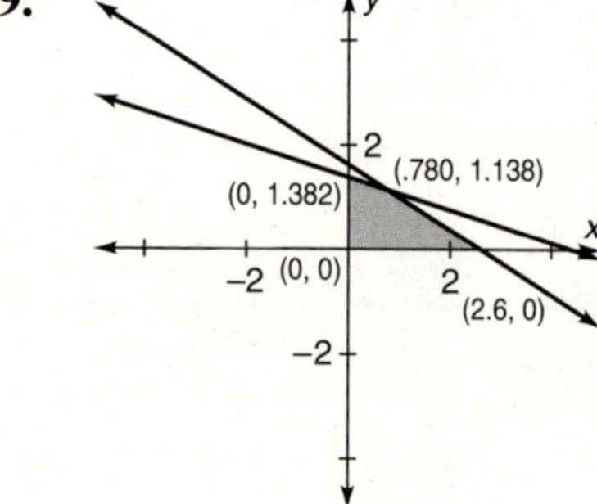
y
2
(.780, 1.138)
(0, 1.382)
x
−2
(0, 0)
2
(2.6, 0)
−2

APPENDIX E

Graphing Calculator Programs for Earth Algebra

The following programs are designed to give the equation of either a line or a parabola which passes the appropriate number of given points, and the error associated with the equation and the given data. Programs provided are for the Casio 7000 and 7700, Sharp 9300, TI–81 and TI–85.

The authors wish to express their gratitude to our calculator consultant, Professor John Kenelly of Clemson University for writing these programs in support of *Earth Algebra*.

E.1 CASIO 7000 & 7700

Line Fit

```
"ENTER 2 PTS
(P,Q)(R,S)"
"P"? → P: "Q"? → Q
"R"? → R: "S"? → S
"NUMBER OTHER PTS"? → K
Lbl 1
"ENTER NEXT PT"
"X"? → X: "Y"? → Y
```

```
(Y – ((S–Q) ÷ (R–P))(X–P) – Q) → E
"ERROR AT PT"
E ◢
Dsz K
Goto 1
```

Curve Fit

```
"ENTER 3 PTS
(P,Q)(R,S)(T,U)"
"P"? → P: "Q"? → Q
"R"? → R: "S"? → S
"T"? → T: "U"? → U
"NUMBER OTHER PTS"? → K
Lbl 1
"ENTER NEXT PT"
"X"? → X: "Y"? → Y
((U – Q) ÷ (T–P)) – ((S–Q) ÷ (R–P)) → Z
Q–(P(S–Q) ÷ (R-P)) + (RPZ ÷ (T–R)) → C
((S–Q) ÷(R–P)) – (Z (R+P) ÷(T-R)) →B
Z ÷ (T–R) → A
(Y – AX² – BX – C) → E
"ERROR AT PT"
E ◢
Dsz K
Goto 1
```

E.2 SHARP 9300

Line Fit

```
Print "enter 1st point"
Input xa
Input ya
Print "enter 2nd point"
Input xb
```

```
Input yb
Print "number other points"
Input k
Label abc
Print "enter next point"
Input x
Input y
error = ((yb – ya) / (xb – xa) * (x – xa) + ya – y)
Print error
k = k – 1
If k > 0 Goto abc
End
```

Curve Fit

```
dim A[3,3]
dim B[3,1]
dim C[3,1]
dim D[1,3]
Print "enter 3 points"
Input xa
A[1,1] = xa^2
A[1,2] = xa
A[1,3] = 1
Input ya
B[1,1] = ya
Input xb
A[2,1] = xb^2
A[2,2] = xb
A[2,3] = 1
Input yb
B[2,1] = yb
Input xc
A[3,1] = xc^2
A[3,2] = xc
A[3,3] = 1
Input yc
```

```
B[3,1] = yc
mat c = 1/ mat A* mat B
Print "number other points"
Input n
Print "enter next point"
Label abc
Input x
D[1,1] = x^2
D[1,2] = x
D[1,3] = 1
Input y
error = (D*C – y)
Print error
n = n – 1
If n > 0 Goto abc
End
```

E.3 TI–81 AND TI–85

TI–81 Line Fit

```
: ClrStat
: Disp "ENTER 2 POINTS"
: Disp "X"
: Input {x} (1)
: Disp "Y"
: Input {y} (1)
: Disp "X"
: Input {x} (2)
: Disp "Y"
: Input {y} (2)
: LinReg
: Disp "NUMBER OTHER POINTS"
: Input N
: Lbl 1
: Disp "ERROR FOR NEXT POINT"
```

```
: Disp "X"
: Input X
: Disp "Y"
: Input Y
: (Y – (a + X)) → E
: Disp "ERROR"
: Disp E
: DS < (N, 1)
: Goto 1
: Stop
```

TI-81 Curve Fit

```
: 3 → Arow
: 3 → Aeol
: 3 → Brow
: 1 → Beol
: 3 → Crow
: 1 → Ceol
: 1 → [A]
: Disp " ENTER 3 PTS"
: 1 → K
: Lbl 1
: Disp "x"
: Input X
: x² → [A] (k, 1)
: x → [A] (k, 2)
: Disp "Y"
: Input [B] (k, 1)
: IS > (k, 3)
: Goto 1
: [A]⁻¹ [B] → [C]
: Disp "NUMBER OTHER PTS?"
: Input N
: Lbl 2
: Disp "ENTER NEXT POINT"
: Disp "X"
```

```
: Input X
: Disp "Y"
: Input Y
: (y – [C] (1, 1)x^2 + [C] (2, 1)x + [C] (3, 1)) → E
: Disp "Error"
: Disp E
: DS < (N , 1)
: Goto 2
: Stop
```

TI - 85 Line Fit

```
:2 → dimL L1
:2 → dimL L2
: Disp " ENTER 2 POINTS"
: For (K, 1, 2, 1)
: Disp "X"
: Input X
: X → L1 (K)
: Disp "Y"
: Input Y
: Y → L2 (K)
: End
: L1 → xStat
: L2 → yStat
: LinR
: Disp " HOW MANY OTHER PTS?"
: Input N
: For (J, 1, N, 1)
: Disp "ENTER NEXT POINT"
: Disp "X"
: Input X
: Disp "Y"
: Input Y
: ERROR = (Y – (a * X + L))
: Disp " ERROR" , ERROR
: End
```

TI-85 Curve Fit

```
: {3,3} → dim A
: {3,1} → dim B
: {3,1} → dim C
: {1,3} → dim D
: Disp "ENTER 3 POINTS"
: For (K, 1, 3, 1)
: Prompt X
: X² → A(K,1)
: X → A(K,2)
: 1 → A(K,3)
: Prompt Y
: Y → B(K,1)
: END
: A⁻¹ * B → C
: Disp "NUMBER OTHER PTS?"
: Prompt N
: For (J, 1, N, 1)
: Disp "ENTER NEXT POINT"
: Prompt X
: X² → D(1,1)
: X → D(1,2)
: 1 → D(1,3)
: Prompt Y
: (det (D*C) – Y) → ERROR
: Disp " ERROR AT THE POINT IS ", ERROR
: END
```

BIBLIOGRAPHY

The references listed here are not only source books used for information in this text, but also include some additional related readings which should be of interest.

Abbey, Edward. *Desert Solitaire* (New York: Simon & Schuster, 1968).

Bates, Albert. *Climate in Crisis: The Greenhouse Effect and What We Can Do* (Summerton, Tenn.: The Book Publishing Company, 1990).

Berry, Wendell. *What are People for?* (Berkeley, California: North Point, 1990).

Brown, Lester. *State of the World 1991: Worldwatch Institute Report on Progress Toward a Sustainable Society* (New York: Norton, 1991)

Ehrlich, Gretel. *The Solace of Open Spaces* (New York: Viking, 1985).

Falk, Jim and Andrew Brownlow. *The Greenhouse Challenge: What's to Be Done?* (Ringwood, Victoria: Penguin Books Australia, 1989).

Foreman, Dave. *Confessions of a Eco-Warrior* (New York: Harmony, 1991).

Gaia. *An Atlas of Planet Management* (New York: Anchor Press/Doubleday, 1984).

Global Tomorrow Coalition. *The Global Ecology Handbook* (Boston: Beacon Press, 1990).

Gribben, John. *Hothouse Earth: The Greenhouse Effect and GAIA* (New York: Grove Weidenfeld, 1990).

Harte, John. *Consider a Spherical Cow: A Course in Environmental Problem Solving* (Mill Valley, CA: University Science Books, 1988).

Lamb, Marjorie. *Two Minutes a Day for a Greener Planet* (New York: HarperCollins, 1991).

Leggett, Jeremy (editor) *Global Warming, the Greenpeace Report* (Oxford: Oxford University Press, 1990).

Lopez, Barry. *The Rediscovery of North America* (Lexington, Ky., University of Kentucky, 1990).

Merwin, W.S. *The Rain in the Trees* (New York: Alfred A. Knopf , 1988).

Mitchell, Finis. *Wind River Trails* (Salt Lake City: Wasatch, 1975).

Nabokov, Ed Peter. *Native America Testimony* (New York: Viking , 1985).

Oliver, Mary. *New and Collected Poems* (Boston: Beacon Press, 1992).

Rich, Adrienne. *An Atlas of the Difficult World* (New York: Norton, 1991).

Roan, Sharon. *Ozone Crisis* (New York: John Wiley, 1989).

The Earth Works Group. *50 Simple Things You Can Do to Save the Earth* (Berkeley, Calif., 1989).

The Student Environmental Action Coalition. *The Student Environmental Action Guide: 25 Simple Things We Can Do* (New York: HarperCollins, 1991).

Tietenberg, Tom. *Environmental and Natural Resource Economics,* 3rd edition (New York: HarperCollins, 1992).

U.S. Bureau of the Census, *Statistical Abstract of the United States* (Washington, D.C.: U.S. Government Printing Office, 1991).

U.S. Bureau of the Census, *Historical Statistics of the United States, Colonial Times to 1970* (Washington, D.C.: U.S. Government Printing Office, 1976).

U.S. Congress, Office of Technology Assessment, *Changing by Degrees: Steps to Reduce Greenhouse Gases* (Washington, D.C.: U.S. Government Printing Office, 1976).

U.S. Department of Energy, Energy Information Administration, *The Motor Gasoline Industry: Past, Present, and Future* (Washington, D.C.: U.S. Government Printing Office, 1991).

Worldwatch Institute, *Worldwatch Papers* (Washington, D.C.).

World Resources Institute, *World Resources 1990–91* (New York, Oxford: Oxford University Press, 1990).

World Resources Institute, *The 1992 Information Please © Environmental Atlas* (Boston: Houghton Mifflin Company, 1992).